Lectures on

STOCHASTIC PROCESSES

By

K. ITO

Notes by

K. MURALIDHARA RAO

Distributed for the

TATA INSTITUTE OF FUNDAMENTAL RESEARCH

SPRINGER - VERLAG

Berlin Heidelberg New York Tokyo

1984

Author

K. ITO
Gakushuin University
Mejiro, Toshimaku
Tokyo
JAPAN

ISBN 3-540-12873-5 Springer-Verlag, Berlin. Heidelberg. New York. Tokyo
ISBN 0-387-12873-5 Springer-Verlag, New York. Heidelberg. Berlin. Tokyo

Printed by M. N. Joshi at the Book Centre Limited, Sion East, Bombay 400 022 and published by the Tata Institute of Fundamental Research, Bombay

PRINTED IN INDIA

PREFACE

In this course of lectures I have discussed the elementary parts of Stochastic Processes from the view point of Markov Processes. I owe much to Professor H. P. McKean's lectures at Kyoto University (1957-58) in the preparation of these lectures.

I would like to express my hearty thanks to Professor K. Chandrasekharan, Dr. K. Balagangadharan, Dr. J.R. Choksi and Mr. K.M. Rao for their friendly aid in preparing the manuscript.

K. Ito

CONTENTS

		Page
Section 0 - Preliminaries		
§1.	Measurable space	1
§2.	Probability space	2
§3.	Independence	5
§4.	Conditional expectation	6
§5.	Wiener and Poisson processes	7
Section 1 - Markov processes		
§1.	Introduction	17
§2.	Definition	20
§3.	Transition probability	21
§4.	Semi-groups	28
§5.	Green operator	30
§6.	The generator	32
§7.	Examples	36
§8.	Dual notions	46
§9.	A theorem of Kac	50
Section 2 - Strong Markov processes		
§1.	Markov time	56
§2.	Examples of Markov time	58
§3.	Definition of strong Markov process ...	59
§4.	A sufficient condition for a Markov process to be strong Markov	60

Page

§5. Example of a Markov process which is not strong Markov 63

§6. Dynkin's formula and generalized first passage time relations 67

§7. Blumenthal's 0-1 law 74

§8. Markov process with discrete state space 75

§9. Generator in the restricted sense ... 80

Section 3 - Multi-dimensional Brownian motion

§1. Definition 90

§2. Generator of the k-dimensional Brownian motion 92

§3. Stochastic solution of Dirichlet problem 96

§4. Recurrence 105

§5. Green function 109

§6. Hitting probability ($k \geqslant 3$) 118

§7. Regular points 126

§8. Plane measure of a two-dimensional Brownian motion curve 129

Section 4 - Additive processes

§1. Definitions 134

§2. Gaussian additive processes and Poisson additive processes 135

§3. Lévy's canonical form 139

§4. Temporally homogeneous Lévy processes ... 153

§5. Stable processes 157

§6. Lévy process as a Markov process ... 162

§7. Multi-dimensional Lévy processes and Riesz Potentials 168

Page

Section 5 - Stochastic differential equations

§ 1. Introduction 172

§ 2. Stochastic integral (1)-Function spaces $\mathcal{E}, \mathcal{L}^2, \mathcal{L}_s$ 175

§ 3. Stochastic integral (2)-Definitions and properties 178

§ 4. Stochastic integral (3)-Continuous version 181

§ 5. Stochastic differentials 187

§ 6. Stochastic differential equations ... 193

§ 7. Construction of diffusion ... 204

Section 6 - Linear diffusion

§ 1. Generalities 211

§ 2. Generator in the restricted sense ... 214

§ 3. Local generator 217

§ 4. Feller's form of generators (1) - Scale 220

§ 5. Feller's form of generators (2) - Speed measure

§ 6. Feller's form of generators (3) 226

§ 7. Feller's form of generators (4) - Conservative compact interval 228

Bibliography 231

Section 0

Preliminaries

§1. Measurable space

Let Ω be a set and let $S(\Omega)$ denote the set of all subsets of Ω. $\mathcal{A} \subset S(\Omega)$ is called an algebra if it is closed under finite unions and complementations; an algebra $\mathcal{B}$ closed under countable unions is called a Borel algebra. For $\mathbb{C} \subset S(\Omega)$ we denote by $\mathcal{A}(\mathbb{C})$ and $\mathcal{B}(\mathbb{C})$, the algebra and Borel algebra, respectively, generated by $\mathbb{C}$. $M \subset S(\Omega)$ is called a monotone class if $A_n \in M$, $n=1,2,\ldots$, and $\{A_n\}$ monotone implies that $\lim_n A_n \in M$. We have the following lemma.

Monotone lemma. If M is a monotone class containing an algebra $\mathcal{A}$ then $M \supset \mathcal{B}(\mathcal{A})$.

The proof of this lemma can be found in P. Halmos: Measure theory.

For any given set Ω we denote by $\mathcal{B}(\Omega)$ a Borel algebra of subsets of Ω.

Definition. A pair $(\Omega, \mathcal{B}(\Omega))$ is called a measurable space. $A \subset \Omega$ is called measurable if $A \in \mathcal{B}(\Omega)$.

Let $(\Omega_1, \mathcal{B}_1(\Omega_1))$ and $(\Omega_2, \mathcal{B}_2(\Omega_2))$ be measurable spaces. A function $f: \Omega_1 \longrightarrow \Omega_2$ is called measurable with respect to $\mathcal{B}_1(\Omega_1)$ if for every $A \in \mathcal{B}_2(\Omega_2)$, $f^{-1}(A) \in \mathcal{B}_1(\Omega_1)$.

Now suppose that Ω_1 is a set, $(\Omega_2, \mathcal{B}(\Omega_2))$ a measurable space and f a function on Ω_1 into Ω_2. Let $\mathcal{B}(f)$ be the class of all sets of the form $f^{-1}(A)$ for $A \in \mathbb{B}_2(\Omega_2)$. Then $\mathbb{B}(f)$ is a Borel algebra and is the least Borel algebra with respect to which f is measurable.

Let $(\Omega_i, \mathbb{B}_i(\Omega_i))$, $i \in I$, be measurable spaces. Let $\Omega = \prod_i \Omega_i$ denote the Cartesian product of Ω_i and let $\pi_i : \Omega \to \Omega_i$ be defined by $\pi_i(w) = w_i$. Let $\mathbb{B}(\Omega)$ be the least Borel algebra with respect to which all the π_i's are measurable. The pair $(\Omega, \mathbb{B}(\Omega))$ is called the <u>product measurable space</u>. $\mathbb{B}(\Omega)$ is the least Borel algebra containing the class of all sets of the form

$$\{f : f_i \in E_i\},$$

where $E_i \in \mathbb{B}_i(\Omega_i)$. A function F into Ω is measurable if and only if $\pi_i F$ is measurable for every $i \in I$.

§2. Probability space

Let Ω be a set, $\mathbb{A} \subset S(\Omega)$ an algebra. A function p on $\mathbb{A}$ such that $p(\Omega) = 1$, $0 \leq p(E) \leq 1$ for $E \in \mathbb{A}$, and such that $p(E \cup F) = p(E) + p(F)$ whenever $E, F \in \mathbb{A}$ and $E \cap F = \phi$, is called an <u>elementary probability measure</u> on $\mathbb{A}$. Let $(\Omega, \mathbb{B}(\Omega))$ be a measurable space and p an elementary probability measure on $\mathbb{B}$. If $A_n \in \mathbb{B}$, A_n disjoint, imply $p(\bigcup_n A_n) = \sum p(A_n)$, we say that p is a <u>probability measure</u> on $\mathbb{B}(\Omega)$. The proof of the following important theorem can be found in P. Halmos : Measure theory.

Theorem (Kolmogoroff)

If p is an elementary probability measure on $\mathcal{A}$ then p can be extended to a probability measure P on $\mathcal{B}(\mathcal{A})$ if and only if the following continuity condition is satisfied:

$$A_n \in \mathcal{A},\ A_n \supset A_{n+1},\ \bigcap_n A_n = \phi \text{ imply } \lim_n p(A_n) = 0.$$

Further under the above condition the extension is unique.

Definition. A triple $(\Omega, \mathcal{B}, P)$, where P is a probability measure on $\mathcal{B}$, is called a probability space.

A real-valued measurable function on a probability space is called a random variable. If a random variable x is integrable we denote the integral by $E(x)$ and call it the expectation of x.

Let $(\Omega_2, \mathcal{B}_2)$ be a measurable space, $(\Omega_1, \mathcal{B}_1, P_1)$ a probability space and $f : \Omega_1 \to \Omega_2$ a measurable function. Define $P_2(E) = P_1(f^{-1}(E))$ for every $E \in \mathcal{B}_2$. Then $(\Omega_2, \mathcal{B}_2, P_2)$ is a probability space and for every integrable function g on Ω_2, $E(g \circ f) = \int g \circ f \, dP_1 = \int g dP_2 = E(g)$. We say that f induces a measure on $\mathcal{B}_2$. In case x is a random variable, the measure induced on the line is called the probability distribution of x.

We shall prove the following formulae which we use later.

Inclusion-exclusion formula. Let $(\Omega, \mathcal{B}, P)$ be a probability space and $A_i \in \mathcal{B}$, $i = 1,2,\ldots, n$. Then

$$P\left(\bigcup_{i=1}^{n} A_i\right) = \sum_i P(A_i) - \sum_{i<j} P(A_i \cap A_j) + \sum_{i<j<k} P(A_i \cap A_j \cap A_k) - \ldots$$

To prove this, let χ_B denote the characteristic function of B . Then

$$P(\cup A_i) = E(\chi_{\cup A_i}) = E(1 - \chi_{\cap A_i^c}) = 1 - E(\Pi \chi_{A_i^c})$$

$$= 1 - E((1 - \chi_{A_1})(1 - \chi_{A_2}) \ldots (1 - \chi_{A_n}))$$

$$= 1 - E\left[1 - \sum_i \chi_{A_i} + \sum_{i<j} \chi_{A_i}\chi_{A_j} - \sum_{i<j<k} \chi_{A_i}\chi_{A_j}\chi_{A_k} + \ldots\right]$$

$$= \sum_i E(\chi_{A_i}) - \sum_{i<j} E(\chi_{A_i \cap A_j}) + \sum_{i<j<k} E(\chi_{A_i \cap A_j \cap A_k}) - \ldots$$

$$= \sum_i P(A_i) - \sum_{i<j} P(A_i \cap A_j) + \ldots$$

The following <u>dual inclusion-exclusion formula</u> is due to Hunt. We have

$$P(\cap A_i) = 1 - P(\cup A_i^c) = 1 - \left\{\sum_i P(A_i^c) - \sum_{i<j} P(A_i^c \cap A_j^c) + \ldots\right\}$$

$$= 1 - \left\{\sum_i (1 - P(A_i)) - \sum_{i<j} (1 - P(A_i \cup A_j)) + \ldots\right\}$$

$$= 1 - \left\{n - \sum_i P(A_i) - \binom{n}{2} + \sum_{i<j} P(A_i \quad A_j) + \ldots\right\}$$

$$= \left[1 - \binom{n}{1} + \binom{n}{2} \ldots\right] + \sum_i P(A_i) - \sum_{i<j} P(A_i \cup A_j) + \ldots$$

$$= \sum_i P(A_i) - \sum_{i<j} P(A_i \cup A_j) + \ldots$$

A collection $(x_t, t \in T)$ of random variables x_t, T being some indexing set, is called a <u>stochastic</u> or <u>random process</u>. We generally assume that the indexing set T is an interval of real numbers.

Let $\{x_t, t \in T\}$ be a stochastic process. For a fixed ω $x_t(\omega)$ is a function on T, called a <u>sample function</u> of the process.

Lastly, an <u>n-dimensional random variable</u> is a measurable function into R^n; an <u>n-dimensional random process</u> is a collection of n-dimensional random variables.

§3. <u>Independence</u>.

Let $(\Omega, \mathcal{B}, P)$ be a probability space and $\mathcal{B}_i$, $i = 1,2,\ldots,n$, n Borel subalgebras of $\mathcal{B}$. They are said to be <u>independent</u> if for any $E_i \in \mathcal{B}_i$, $1 \leq i \leq n$, $P(E_1 \cap \ldots \cap E_n) = P(E_1)\ldots P(E_n)$. A collection $(\mathcal{B}_\alpha)_{\alpha \in I}$ of Borel subalgebras of $\mathcal{B}$ is said to be independent if every finite subcollection is independent.

Let $x_1,\ldots,x_n$ be n random variables on $(\Omega, \mathcal{B}, P)$ and $\mathcal{B}_i$, $1 \leq i \leq n$, the least Borel subalgebra of $\mathcal{B}$ with respect to which x_i is measurable. $x_1,\ldots,x_n$ are said to be independent if $\mathcal{B}_1,\ldots,\mathcal{B}_n$ are independent.

Finally, suppose that $\{x_\alpha(t,w)\}_{\alpha \in I}$ is a system of random processes on Ω and $\mathcal{B}_\alpha$ the least Borel subalgebra of $\mathcal{B}$ with respect to which $x_\alpha(t,w)$ is measurable for all t. The processes are said to be <u>stochastically independent</u> if the $\mathcal{B}_\alpha$ are independent.

We give some important facts about independence. If x and y are random variables on Ω the following statements are equivalent:

1) $E(e^{i\alpha x + i\beta y}) = E(e^{i\alpha x})E(e^{i\beta y})$, α and β real;

2) The measure induced by $z(w) = (x(w), y(w))$ on the plane is the product of the measures induced by x and y on the line;

3) x and y are independent.

§ 4. Conditional expectation.

Let $(\Omega, \mathcal{B}, P)$ be a probability space and $\mathbb{C}$ a Borel subalgebra of $\mathcal{B}$. Let $x(w)$ be a real-valued integrable function. We follow Doob in the definition of the conditional expectation of x.

Consider the set function μ on $\mathbb{C}$ defined by $\mu(C) = E(x:C)$. Then $\mu(C)$ is a bounded signed measure and $\mu(C) = 0$ if $P(C) = 0$. Therefore by the Radon-Nikodym theorem there exists a unique (upto P-measure 0) function $\varphi(w)$ measurable with respect to $\mathbb{C}$ such that

$$\mu(C) = E(\varphi : C).$$

Definition. $\varphi(w)$ is called the conditional expectation of x with respect to $\mathbb{C}$ and is denoted by $E(x/\mathbb{C})$.

The conditional expectation is not a random variable but a set of random variables which are equal to each other except for a set of P-measure zero. Each of these random variables is called a version of $E(x/\mathbb{C})$.

The following conclusions (which are valid with probability 1) result from the definition.

1. $E(1/\mathcal{C}) = 1$.

2. $E(x/\mathcal{C}) \geqslant 0$ if $x \geqslant 0$.

3. $E(\alpha x + \beta y/\mathcal{C}) = \alpha E(x/\mathcal{C}) + \beta E(y/\mathcal{C})$.

4. $|E(x/\mathcal{C})| \leq E(|x|/\mathcal{C})$.

5. If $x_n \longrightarrow x$, $|x_n| \leq S$ with $E(S) < \infty$, then

$$\lim_n E(x_n/\mathcal{C}) = E(x/\mathcal{C}).$$

6. If $\sum_n E(|x_n|) < \infty$, then $E(\sum_n x_n/\mathcal{C}) = \sum_n E(x_n/\mathcal{C})$.

7. If x is $\mathcal{C}$- measurable, then $E(xy/\mathcal{C}) = xE(y/\mathcal{C})$.
   In particular, if x is $\mathcal{C}$-measurable, then $E(x/\mathcal{C}) = x$.

8. If x and $\mathcal{C}$ are independent, then $E(x/\mathcal{C}) = E(x)$.

9. If $\mathcal{C} = \{A : P(A) = 0 \text{ or } 1\}$, then $E(x/\mathcal{C}) = E(x)$.

10. If $\mathcal{C}_1 \supset \mathcal{C}_2$, then $E(x/\mathcal{C}_2) = E(E(x/\mathcal{C}_1)/\mathcal{C}_2)$ and, in particular, $E(E(x/\mathcal{C})) = E(x)$.

§5. Wiener and Poisson processes.

The following processes are very important and we shall encounter many examples of these.

We shall define a Wiener process and establish its existence.

Let $\{x_t(w)\ ,\ o \leq t < \infty\}$ be a stochastic process such that

1) for almost all w the sample function $x_t(w)$ is a continuous function on $[o, \infty)$ and vanishes at $t = o$;

2) $P(w : x_{t_1}(w) \in E_1, \ldots, x_{t_n}(w) - x_{t_{n-1}}(w) \in E_n) = P(w : x_{t_1}(w) \in E_1) \ldots$ $\ldots P(w : x_{t_n}(w) - x_{t_{n-1}}(w) \in E_n)$, where $t_1 < t_2 < \ldots < t_n$. This means

that x_{t_1}, $x_{t_2} - x_{t_1}$, ..., $x_{t_n} - x_{t_{n-1}}$ are independent if $t_1 < \ldots < t_n$;

3) $P(w : x_t(w) - x_s(w) \in E) = [2\pi(t-s)]^{-\frac{1}{2}} \int_E e^{-x^2/2(t-s)} dx.$

Then the process is called a <u>Wiener process</u>. This process is extremely important and we shall now construct a Wiener process which we shall use later. This incidentally will establish the existence of Wiener process.

Let $\Omega = C[o, \infty)$ be the space of all real continuous functions on $[o, \infty)$. We introduce an elementary probability measure on Ω as follows.

For any integer n, $o < t_1 < t_2 \ldots < t_n < \infty$ and a Borel set B^n in R^n, let

$$E = \left\{ w : w \in \Omega \text{ and } (w(t_1), \ldots, w(t_n)) \in B^n \right\},$$

and $$p_{t_1 \ldots t_n}(E) = \int \cdots \int_{B^n} N(t_1, o, x_1) N(t_2 - t_1, x_1, x_2) \ldots N(t_n - t_{n-1}, x_{n-1}, x_n) \, dx_1 \ldots dx_n$$

where $$N(t,x,y) = \frac{1}{\sqrt{2\pi t}} e^{-(y-x)^2/2t}$$

If $o < u_1 < \ldots < u_m < \infty$ is a set of points containing $t_1, \ldots, t_n$ and $t_r = u_{i_r}$, $r = 1, 2, \ldots, n$, then E can also be written as

$$E = \left\{ w : w \in \Omega \text{ and } (w(u_1), \ldots, w(u_m)) \in B^m \right\},$$

and then

$$p_{u_1 \dots u_m}(E) = \int_{B^m}\!\!\cdot\!\int N(u_1,o,x_1)\dots N(u_m-u_{m-1},x_{m-1},x_m)dx_1\dots dx_m ,$$

where B^m is the inverse image of B^n under the mapping $(x_1,\dots,x_m) \longrightarrow (x_{i_1},\dots,x_{i_n})$ of R^m into R^n. Using the formula

$$\int N(t,x,y)N(s,y,z)dy = N(t+s,x,z) ,$$

we can show that $p_{u_1..u_m}(E) = p_{t_1..t_n}(E)$.

Now suppose that E has two representations

$$E = \left\{w : (w(t_1),\dots,w(t_n)) \in B^n,\ B^n \subset R^n\right\}$$

$$= \left\{w : (w(s_1),\dots,w(s_m)) \in B^m,\ B^m \subset R^m\right\},$$

and $o < u_1 < \dots < u_r$ is the union of the sets $\{t_1,\dots,t_n\}$ and $\{s_1,\dots,s_m\}$. Then from the above, $p_{t_1..t_n}(E)=p_{u_1..u_r}(E)=p_{s_1..s_m}(E)$. Hence $p_{t_1..t_n}(E)$ does not depend on the choice of the representation for E. We denote this by $p(E)$.

The class $\mathcal{A}$ of all such sets E, for all n, for all such n-tuples $(t_1,\dots,t_n)$ and all Borel sets of R^n, is easily shown to be an algebra. It is not difficult to show that p is an elementary probability measure on $\mathcal{A}$. This elementary probability measure is called the <u>elementary Wiener measure</u>.

We shall presently prove that p satisfies the continuity condition of Kolmogoroff's theorem. Hence p can be extended to a probability measure P on $\mathcal{B}(\mathcal{A})$, which we call the Wiener measure on $(\Omega, \mathcal{B}(\mathcal{A}))$. It will then follow that $P(w : w(o) = o) = 1$.

Now let $x_t(w) = w(t)$. Then evidently $\{x_t, o \leq t < \infty\}$ is a stochastic process with almost all sample functions continuous and vanishing at $t = 0$. We show that $\{x_t, o \leq t < \infty\}$ is a Wiener process.

The function $f : (x_1, x_2) \longrightarrow x_2 - x_1$ of $R^2 \longrightarrow R^1$ is continuous and hence for any Borel set $E \subset R^1$, the set $B = f^{-1}(E)$ $= \{(x_1, x_2) : x_2 - x_1 \in E\}$ is a Borel set in R^2. Therefore

$$P\{w : x_t - x_s \in E\} = P\{w : (w(s), w(t)) \in B = f^{-1}(E)\}$$

$$= \iint_B N(s,o,x_1)N(t-s,x_1,x_2)dx_1dx_2 .$$

The transformation $(x_1, x_2) \longrightarrow (x,y)$ with $x = x_1$, $y = x_2 - x_1$ gives

$$P(w : x_t - x_s \in E) = \iint_{\{(x,y):y\in E\}} N(s,o,x)N(t-s,x,y+x)\,dxdy$$

$$= \int_E N(t-s,o,y)dy.$$

Again

$$P\{w : x_{t_1} \in E_1, \ldots, x_{t_n} - x_{t_{n-1}} \in E_n\} = P\{w : (w(t_1), \ldots, w(t_n)) \in B^n\},$$

where $B^n = \{(x_1, \ldots, x_n) : x_1 \in E_1, x_2 - x_1 \in E_2, \ldots, x_n - x_{n-1} \in E_n\}$.

Therefore

$$P\{w: x_{t_1} \in E_1, \ldots, x_{t_n} - x_{t_{n-1}} \in E_n\} = \int_{B^n}\!\!\cdots\!\!\int N(t_1, o, x_1) \ldots N(t_n - t_{n-1}, x_{n-1}, x_n)\, dx_1 \ldots dx_n$$

$$= \int_{E_1 \times \cdots \times E_n}\!\!\cdots\!\!\int N(t_1, o, x_1') \ldots N(t_n - t_{n-1}, o, x_n')\, dx_1' \ldots dx_n'$$

$$= P\{w: x_{t_1} \in E_1\}\, P\{w: x_{t_2} - x_{t_1} \in E_2\} \ldots P\{w: x_{t_n} - x_{t_{n-1}} \in E_n\},$$

where $x_1' = x_1$, $x_i' = x_i - x_{i-1}$ $i = 2, \ldots, n$. We have proved that (x_t) is a Wiener process.

It remains to prove that p satisfies the continuity condition. We shall prove the following more general theorem.

Theorem (Prohorov, 'Convergence of stochastic processes and limit theorems in Probability Theory', Teoria veroyatnesteii e eyo primenania Vol. I Part 2, 1956).

Let p be an elementary probability measure on $\mathcal{A}$ which is a probability measure when restricted to sets of $\mathcal{A}$ dependent on a fixed set $t_1, \ldots, t_n$. Let E denote expectations with respect to p. If there exist $a > o$, $b > 1$ and $C > o$ such that $E(|x_t - x_s|^a) \leq C|t-s|^b$ then p can be extended to a probability measure on $\mathcal{B}(\mathcal{A})$.

Proof. Let $A_n \supset A_{n+1}$, $n = 1, 2, \ldots, A_n \in \mathcal{A}$ be such that $p(A_n) > \epsilon > o$, for all n. We prove that $\bigcap_n A_n \neq \phi$.

Let $A_n = \{w : (w(t_1^{(n)}), \ldots, w(t_{r_n}^{(n)})) \in B_n\}$, where $B_n \in \mathcal{B}(R^{r_n})$ (the set of Borel subsets of R^{r_n}). For each n there exists a q_n such that a) each $t_i^{(n)} \leq q_n$, b) at most one $t_i^{(n)}$. is contained in any closed interval $[(k-1)2^{-q_n}, k2^{-q_n}]$ for $k=1,2,\ldots,q_n 2^{q_n}$. By adding superfluous suffixes if necessary, one can assume that each point $k\,2^{-q_n}$, $k = 0,1,\ldots,q_n 2^{q_n}$, is in $\{t_1^{(n)},\ldots,t_{r_n}^{(n)}\}$, and moreover (by adding, say, the midpoint if necessary) that in each open interval $((k-1)2^{-q_n}, k2^{-q_n})$ there is exactly one point of $(t_1^{(n)},\ldots,t_{r_n}^{(n)})$. Thus $r_n = q_n 2^{q_n+1}$ and $t_{2k}^{(n)} = k2^{-q_n}$. Finally, by adding superfluous sets when necessary one may assume that $q_n = n$ i.e., that

$$A_n = \left\{ w : (w(t_1^{(n)}),\ldots,w(t_{n2^{n+1}}^{(n)})) \in B_n \right\} ,$$

where $t_{2k}^{(n)} = k2^{-n}$ and $(t_1^{(n)},\ldots,t_{n2^{n+1}}^{(n)}) \subset (t_1^{(n+1)},\ldots,t_{(n+1)2^{n+2}}^{(n+1)})$.

Since p is a probability measure when restricted to sets dependent on a fixed set $s_1,\ldots,s_k$, we can further assume that each B_n is a closed bounded subset of $R^{n2^{n+1}}$. Now, since $E(|x(s)-x(t)|^a) \leq C|s-t|^b$,

$$p(w: |w(t_i^{(n)})-w(t_{i-1}^{(n)})| \geq |t_i^{(n)}-t_{i-1}^{(n)}|^{\delta}) = p(w: |w(t_i^{(n)})-w(t_{i-1}^{(n)})|^a \geq$$

$$\geq |t_i^{(n)}-t_{i-1}^{(n)}|^{a\delta}) \leq C|t_i^{(n)}-t_{i-1}^{(n)}|^{b-a\delta}$$

Choose $\delta > 0$ such that $\lambda = b-a\delta-1 > 0$. Then

$$p(w: |w(t_i^{(n)})-w(t_{i-1}^{(n)})| \geq |t_i^{(n)}-t_{i-1}^{(n)}|^{\delta}) \leq C|t_i^{(n)}-t_{i-1}^{(n)}|^{1+\lambda} \leq C\,2^{-n(1+\lambda)}$$

Hence $p\left(\bigcup_{i=2}^{n2^{n+1}} (w: |w(t_i^{(n)})-w(t_{i-1}^{(n)})| \geq |t_i^{(n)}-t_{i-1}^{(n)}|^{\delta})\right) \leq \frac{Cn.2^{n+1}}{2^{n(1+\lambda)}} = 2\,C\,n.2^{-\lambda n}$.

Since $\sum n2^{-n\lambda}$ is convergent, there exists m_o such that $2C\sum_{n=m_o}^{\infty} n2^{-n\lambda} < \frac{\epsilon}{2}$.

Then for $\ell \geq m_o$,

$$p\left(\bigcup_{n=m_o}^{\ell} \bigcup_{i=2}^{n\cdot 2^{n+1}} (w: |w(t_i^{(n)})-w(t_{i-1}^{(n)})| \geq |t_i^{(n)}-t_{i-1}^{(n)}|^{\delta})\right) < \frac{\epsilon}{2} ,$$

and so

$$p\left(\bigcap_{n=m_o}^{\ell} \bigcap_{i=2}^{n\cdot 2^{n+1}} (w: |w(t_i^{(n)})-w(t_{i-1}^{(n)})| < |t_i^{(n)}-t_{i-1}^{(n)}|^{\delta})\right) > 1 - \frac{\epsilon}{2} .$$

It follows that

$$p\left(A_\ell \cap \bigcap_{n=m_o}^{\ell} \bigcap_{i=2}^{n\cdot 2^{n+1}} (w: |w(t_i^{(n)})-w(t_i^{(n)})| < |t_i^{(n)}-t_{i-1}^{(n)}|^{\delta})\right) > \frac{\epsilon}{2} ,$$

and so this set is non-empty. Call this set B_ℓ'. Then $B_\ell' \supset B_{\ell+1}'$ and $A_\ell \supset B_\ell'$. We prove that $\bigcap_{m_o}^{\infty} B_\ell' \neq \emptyset$.

From each B_ℓ', choose a function w_ℓ linear in each inverval $[t_{i-1}^{(\ell)}, t_i^{(\ell)}]$. Such a function exists, since for each $w \in B_\ell'$ there corresponds such a function determined completely by $(w(t_1^{(\ell)}),\ldots,w(t_{\ell\cdot 2^{\ell+1}}^{(\ell)}))$. We can assume that $w(t_1^{(\ell)}) = 0$, since zero never occurs in the points which define sets of $\mathcal{A}$. Now if $\ell \geq m_o$,

$$|w_\ell(t_i^{(n)})-w_\ell(t_{i-1}^{(n)})| < |t_i^{(n)}-t_{i-1}^{(n)}|^{\delta} \leq 2^{-n\delta} , \quad m_o \leq n \leq \ell , \quad 1 \leq i \leq n2^{n+1} ,$$

so that $|w_\ell(k2^{-n})-w_\ell((k-1)2^{-n})| \leq 2.2^{-n\delta}$, $1 \leq k \leq n2^n$, $m_o \leq n \leq \ell$. Given $k2^{-\ell}$, $k'2^{-\ell}$, $k' < k$, $k2^{-\ell} - k'2^{-\ell} < 2^{-m_o}$, there exists $q \leq \ell$ such that $2^{-q} \leq k2^{-\ell} - k'2^{-\ell} < 2^{-q+1}$. In the interval $[k'2^{-\ell}, k2^{-\ell}]$ there exist at most two points of the form $j2^{-q}$, $(j+1)2^{-q}$. Then since $w_\ell \in B'_q$, $|w_\ell(j2^{-q})-w_\ell((j+1)2^{-q})| < 2.2^{-q\delta}$. Repeating similar arguments, we can prove that

$$|w_l(k2^{-\ell}) - w_\ell(k'2^{-\ell})| \leq 4(1-2^{-\delta})^{-1}2^{-q\delta} \leq \lambda|k2^{-\ell}-k'2^{-\ell}|^{\delta} ,$$

λ being a constant. Now we can easily see that

$$|w_\ell(t_i^{(\ell)})-w_\ell(t_j^{(\ell)})| < \mu|t_i^{(\ell)}-t_j^{(\ell)}|^{\delta} \quad \text{if } |t_i^{(\ell)}-t_j^{(\ell)}| \leq 2^{-m_o} \quad \text{say.}$$

From this it easily follows, using linearity of w_ℓ in each interval $[t_i^{(\ell)}, t_{i+1}^{(\ell)}]$, that if $t_i^{(\ell)} \leq t \leq s \leq t_j^{(\ell)}$, then

$$|w_\ell(t) - w_\ell(s)| \leq 4\mu|t_i^{(\ell)}-t_j^{(\ell)}|^{\delta} .$$

Now since $w_{\ell+p} \in A_\ell$ for every $p \geq o$, $(w_{\ell+p}(t_1^{(\ell)}),\ldots,w_{\ell+p}(t_{\ell 2^{\ell+1}}^{(\ell)})) \in B_\ell$. Since B_ℓ is compact, this sequence has a limit point in B_ℓ. Since the same is true for every ℓ , we can, by the diagonal method, extract a subsequence $\{w_n\}$, say, such that $w_n(t_i^{(\ell)})$ converges for all i and for all ℓ .

Now let t_o and $\eta > o$ be given. For large n_o , suppose that $t_i^{(n_0)} \leq t_0 \leq t_{i+1}^{(n_0)}$, $|t_i^{(n_0)} - t_{i+1}^{(n_0)}| < 2^{-n_0} < \frac{\eta}{2}$. Then if ℓ and m are large and $t_i^{(n_0)} \leq t_j^{(\ell)} \leq t_0 \leq t_{j+1}^{(\ell)} \leq t_{i+1}^{(n_0)}$,

$t_i^{(n_0)} \leq t_k^{(m)} \leq t_0 \leq t_{k+1}^{(m)} \leq t_{i+1}^{(n_0)}$, we have

$$|w_\ell(t_0)-w_m(t_0)| \leq |w_\ell(t_0)-w_\ell(t_j^{(\ell)})| + |w_\ell(t_j^{(\ell)})-w_\ell(t_i^{(n_0)})| + |w_\ell(t_i^{(n_0)})-w_m(t_i^{(n_0)})| +$$

$$+ |w_m(t_i^{(n_0)})-w_m(t_k^{(m)})| + |w_m(t_k^{(m)})-w_m(t_0)|$$

$$\leq |t_0-t_j^{(\ell)}|^\delta + \mu |t_j^{(\ell)}-t_i^{(n_0)}|^\delta + \frac{\eta}{2} + |t_i^{(n_0)}-t_k^{(m)}|^\delta \mu + |t_k^{(m)}-t_0|^\delta < A\frac{\eta}{2} ,$$

A being some constant. This is true for any $t \in [t_i^{(n_0)}, t_{i+1}^{(n_0)}]$. This shows that the limit exists at every point of R^1. Also using $|w_\ell(t)-w_\ell(s)| < 4\mu|t_i^{(\ell)}-t_j^{(\ell)}|^\delta$, we easily see that the limit function, say w , is continuous. Also since $(w(t_1^{(\ell)}),\dots,w(t_{\ell 2^\ell+1}^{(\ell)})) \in B_\ell$ for all ℓ , $\bigcap_{\ell \geqslant m_0} B'_\ell \neq \emptyset$. We have proved the theorem.

In our case we have

$$p(x_t - x_s \in E) = [2\pi(t-s)]^{-\frac{1}{2}} \int_E e^{\frac{-x^2}{2(t-s)}} dx$$

and thus $E(|x_t - x_s|^4) = 3(t-s)^2$.

<u>Poisson processes</u>. Let $(x_t , 0 \leq t < \infty)$ be a stochastic process such

that

1. for almost all w the sample function $x_t(w)$ is a step function increasing with jump 1 and vanishes at $t = o$;

2. $P(x_t - x_s = k) = e^{-\lambda(t-s)} \dfrac{(t-s)^k \lambda^k}{k!}$ with $\lambda > o$;

3. $P(x_{t_1} \in E_1, x_{t_2} - x_{t_1} \in E_2, \ldots, x_{t_n} - x_{t_{n-1}} \in E_n) = P(x_{t_1} \in E_1) \ldots$

$\ldots P(x_{t_n} - x_{t_{n-1}} \in E_n)$; i.e. $x_{t_1}, \ldots, x_{t_n} - x_{t_{n-1}}$ are independent if $t_1 < t_2 \; \ldots . < t_n$; then the process is called a Poisson process.

Section 1

Markov processes

§1. Introduction

In the following lectures, we shall be mainly concerned with Markov processes, and in particular with diffusion processes.

We shall first give an intuitive explanation and then a mathematical definition. The intuitive model of a Markov process is a phenomenon changing with time according to a certain stochastic rule and admitting the possibility of a complete stop. The space of the Markov process has the set of possible states of the phenomenon as its counterpart in the intuitive model. Specifically, consider a moving particle. Its possible positions are points of a space S and its motion is governed by a stochastic rule. The particle may possibly disappear at some time; we then say it has gone to its <u>death point</u>. A possible motion is a mapping of $[0, \infty)$ into the space of positions. Such a function is a <u>sample path</u>. The set of all sample paths is the <u>sample space</u> of the process, denoted by W. A probability law P_a governing the path of the particle starting at a point $a \in S$ is a probability distribution on a Borel algebra of sub-sets of W. The stochastic rule consists of a system of probability laws governing the path. Finally, the condition on the system, that "if the particle arrives at a position 'a' at time 't' it starts afresh according to the probability law P_a ignoring its past history" will correspond intuitively to the basic <u>Markov property</u>.

<u>Definitions</u>. We turn now to the mathematical definitions. We first explain the notation and terminology which we shall use.

Let S denote a locally compact Hausdorff space satisfying the second axiom of countability. Let $\mathbb{B}(S)$ denote the set of all Borel sub-sets of S, $\mathcal{B}(S)$ the set of all $\mathbb{B}(S)$-measurable bounded functions on S. Since S satisfies the second axiom of countability, this class coincides with the class of all bounded Baire functions on S. We shall add to S a point ∞ to get a space $S \cup \{\infty\}$. $S \cup \{\infty\}$ has the topology which makes S an open sub-space and ∞ an isolated point. Then if $\mathbb{B}(S \cup \{\infty\})$ and $\mathcal{B}(S \cup \{\infty\})$ are defined in the same way, $\mathbb{B}(S) \subseteq \mathbb{B}(S \cup \{\infty\})$, and for any $f \in \mathcal{B}(S)$ if we put $f(\infty) = 0$, then $f \in \mathcal{B}(S \cup \{\infty\})$. A function $w : [0,\infty] \longrightarrow S \cup \{\infty\}$ is called a <u>sample path</u> if

1) $w(\infty) = \infty$;

2) there exists a number $\sigma_\infty(w) \in [0,\infty]$ such that $w(t) = \infty$ for $t \geqslant \sigma_\infty(w)$ and $w(t) \in S$ for $t < \sigma_\infty(w)$;

3) $w(t)$ is right continuous for $t < \sigma_\infty(w)$.

For any sample path w, $\sigma_\infty(w)$ is called the <u>killing time</u> of the path. For any path w we denote by $x_t(w)$ the value of w at t, i.e., $x_t(w) = w(t)$. Then we can regard x as a function of the pair (t,w). Given a sample path w the paths w_s^- and w_s^+ defined for any s by

$$x_t(w_s^-) = x_{t \wedge s}(w) \quad \text{if} \quad t < \infty ,$$

and

$$x_{\infty}(w_s^-) = \infty ,$$

where

$$t \wedge s = \min(t,s) ;$$

$$x_t(w_s^+) = x_{t+s}(w) ,$$

are called the stopped path and the shifted path at time s, respectively. A system W of sample paths is called a sample space if $w \in W$ implies $w_s^- \in W$, $w_s^+ \in W$ for each s. For a sample space W the Borel algebra generated by sets of the form $(w: w \in W,\ x_t(w) \in E)$, $t \in [0, \infty)$, $E \in \mathcal{B}(S)$ is denoted by $\mathcal{B}$ or $\mathcal{B}(W)$, and $\mathfrak{B}$ or $\mathfrak{B}(W)$ denotes the set of all bounded $\mathcal{B}$-measurable functions on W. The class of all sets of the form $(w: w_s^- \in B)$, $B \in \mathcal{B}$, is called the stopped Borel algebra at s, and is denoted by $\mathcal{B}_s$ or $\mathcal{B}_s(W)$. $\mathfrak{B}_s$ will denote the system of all bounded $\mathcal{B}_s$-measurable functions, Note that $\mathcal{B}_s$ increases with s and $\mathcal{B}_{\infty} = \mathcal{B}$.

Consider the function $x(t,w)$ on $R \times W$ into $S \cup \{\infty\}$. Let

$$x_n(t,w) = x(\frac{j+1}{2^n}, w) = w(\frac{j+1}{2^n}) \quad \text{for} \quad \frac{j}{2^n} < t \leq \frac{j+1}{2^n} .$$

Then $x_n(t,w)$ is measurable with respect to $\mathcal{B}(R) \times \mathcal{B}(W)$ and $x_n(t,w) \longrightarrow x(t,w)$ pointwise. $x_t(w)$ is therefore a measurable function of the pair (t,w).

§2. Definition. A Markov process is a triple

$$\mathcal{M} = (S, W, P_a, a \in S \cup \{\infty\})$$

where

1) S is a locally compact Hausdorff space with the second axiom of countability;

2) W is a sample space;

3) $P_a(B)$ are probability laws for $a \in S \cup \{\infty\}$, $B \in \mathcal{B}$, i.e.

   a) $P_a(B)$ is a probability measure in $\mathcal{B}$ for every $a \in S \cup \{\infty\}$,

   b) $P_a(B)$, for fixed B, is $\mathcal{B}(S)$-measurable in a,

   c) $P_a(x_o = a) = 1$,

   d) P_a has the Markov property i.e.

   $B_1 \in \mathcal{B}_t$, $B_2 \in \mathcal{B}$ imply

$$P_a\left[w: w \in B_1, w_t^+ \in B_2\right] = E_a\left[w \in B_1 : P_{x_t}(B_2)\right],$$

where the second member is by definition equal to $\int_{B_1} P_{x_t(w)}(B_2)\,dP_a$.

[For fixed $t, B_2, P_{x_t(w)}(B_2)$ is a bounded measurable function on W.]

Remark 1. d) is equivalent to the following:

$$f \in \mathcal{B}_t,\ g \in \mathcal{B} \text{ imply } \int f(w)g(w_t^+)\,dP_a = E_a\left[f(w)E_{x_t(w)}(g(w'))\right]$$

More generally d) is equivalent to

d') $f \in \mathcal{B}_t$, $g \in \mathcal{B}$, $B_1 \in \mathbb{B}_t$, $B_2 \in \mathbb{B}$ imply

$$E_a\left[f(w)g(w_t^+) : w \in B_1, w_t^+ \in B_2\right] = E_a\left[w \in B_1 : f(w) E_{x_t(w)}(w' \in B_2 : g(w'))\right].$$

S , W , P_a are called the state space, sample space and probability law of the process respectively.

We give below three important examples of the sample space in a Markov process.

a) $W = W_{rc}$ = the set of all sample paths.

These processes are called right continuous Markov processes.

b) $W = W_{d_1}$ = the set of sample paths whose only discontinuities before the killing time are of the first kind, i.e. $w(t-0)$, $w(t+0)$ exist and $w(t-0) \neq w(t+0) = w(t)$, $t < \sigma_\infty(w)$. These are called Markov processes of type d_1.

c) $W = W_c$ = the set of all sample paths which are continuous before the killing time. These are continuous Markov processes.

Remark 2. A Markov process is called conservative if $P_a(\sigma_\infty = \infty) = 1$ for all a.

§3. Transition Probability

The function $P(t,a,E) = P_a(x_t \in E)$ on $\mathbb{B}(S)$, $a \in S$ and $0 < t < \infty$ being fixed, is a measure on $\mathbb{B}(S)$ called the transition probability of P_a at time t. The transition probability has the following properties:

(T.1) $P(t,a,E)$ is a sub-stochastic measure in E , i.e. it is a measure in E with total measure $\leq$ 1.

For $P(t,a,S) = P_a(x_t \in S) = 1 - P_a(x_t = \infty) \leq 1$.

(T.2) $P(t,a,E) \in \mathcal{B}(S)$ for fixed t and E.

For $P(\ ,a,E) = P_a(B)$ where $B = \{x_t \in E\}$ and $P_a(B)$ is by definition $\mathcal{B}(S)$-measurable in a for fixed B.

(T.3) $P(t,a,E)$ is measurable in the pair (t,a) for fixed E.

For $f \in \mathcal{B}(S)$ let

$$H_t(f(a)) = \int_S P(t,a,db)f(b) = \int_{W-\{x_t=\infty\}} f(w(t))dP_a .$$

$$= \int_W f(w(t))dP_a , \text{ since } f(\infty) = 0 .$$

If f is a bounded continuous function and $\delta_n \downarrow 0$

$$\lim_{\delta_n \to 0} H_{t+\delta_n} f(a) = \lim_{\delta_n \to 0} \int_W f(w(t+\delta_n))dP_a .$$

$$= \int_W f(\lim_{\delta_n \to 0} w(t+\delta_n))dP_a = \int_W f(w(t))dP_a ,$$

since $w(t)$ is right continuous.

$H_t f(a)$ is thus right continuous in t, if f is bounded and continuous. It is not difficult to show (by considering simple functions and then generalizing) that $H_t f(a)$ is measurable in a if

f is measurable. Therefore $H_t f(a)$ is measurable in the pair (t,a) if f is continuous and bounded. Further, if $\{f_n\}$ is a sequence of measurable functions with $|f_n| \leq \eta$ and $f_n \longrightarrow f$, then $H_t f_n \longrightarrow H_t f$. The class of those measurable functions f for which $H_t f(a)$ is measurable in the pair (t,a) thus contains bounded continuous functions and is closed for limits. Therefore $H_t f(a)$ is measurable in the pair (t,a) for $f \in \mathcal{B}(S)$. If $f = \chi_E$, $H_t f(a) = P(t,a,E)$.

(T.4) $\lim_{t \downarrow 0} P(t,a,U_a) = 1$, where U_a is an open set containing a.

Let $t_n \downarrow 0$, and $B_n = \{ w : w(t_n) \in U_a \}$. Since $w(t)$ is right continuous, $\{w : w(0) \in U_a\} \subset \bigcup_{n=1}^{\infty} \bigcap_{m=n}^{\infty} B_m$.

Therefore $\liminf_{t_n \downarrow 0} P(t_n, a, U_a) \geqslant P_a\left[\bigcup_{n=1}^{\infty} \bigcap_{m=n}^{\infty} B_m \right]$

$$\geqslant P_a\{w : w(0) \in U_a\} \geqslant P_a\{w : w(0) = a\} = 1.$$

(T.5) <u>Chapman-Kolmogoroff equation</u>:

$$P(t+s,a,E) = \int_S P(t,a,db)P(s,b,E).$$

$$P(t+s,a,E) = P_a\{x_{t+s} \in E\} = P_a\{x_t \in S , x_{t+s} \in E\}$$

$$= P_a\{x_t \in S , x_s(w_t^+) \in E\}$$

$$= E_a\left[x_t \in S : P_{x_t}\{x_s(w) \in E\}\right]$$

$$= E_a\left[x_t \in S : P(s,x_t,E)\right]$$

$$= \int_S P(t,a,db)\, P(s,b,E)$$

(T.6) $P_a(x_{t_1} \in E_1, \ldots, x_{t_n} \in E_n)$.

$$= \int_{a_i \in E_i} \int P(t_1,a,da_1) P(t_2-t_1,a_1,da_2) \ldots P(t_n-t_{n-1},a_{n-1},da_n)$$

$(0 \leq t_1 < t_2 < \cdots < t_n)$. **We shall prove this for $n = 2$.**

$$P_a(x_{t_1} \in E_1, x_{t_2} \in E_2) = P_a(x_{t_1} \in E_1, x_{(t_2-t_1)+t_1} \in E_2)$$

$$= P_a(x_{t_1} \in E_1 , x_{t_2-t_1}(w^+_{t_1}) \in E_2)$$

$$= P_a(w \in B_1 , w^+_{t_1} \in B_2) , \quad B_1 = \{x_{t_1} \in E_1\} \text{ and }$$

$$B_2 = \{x_{t_2-t_1} \in E_2\}$$

$$= \int_{B_1} P_{x_{t_1}}(B_2) dP_a = \int_{B_1} P_{x_{t_1}}(w : w(t_2-t_1) \in E_2) dP_a$$

$$= \int_{E_1} P(t_1,a,da_1) P(t_2-t_1,a_1,E_2)$$

$$= \iint_{a_i \in E_i} P(t_1,a,da_1) P(t_2-t_1,a_1,da_2).$$

(T.7) Suppose that $\mathcal{M}_1 = (S_1, W_1 P^1_a , a \in S_1 \cup \{\infty\})$ and $\mathcal{M}_2 = (S_2, W_2, P^2_a , a \in S_2 \cup \{\infty\})$ are two Markov processes with

$S_1 = S_2$, $W_1 = W_2$ and $P^1(t,a,E) = P^2(t,a,E)$; then $\mathcal{M}_1 \equiv \mathcal{M}_2$,
i.e. $P_a^1 = P_a^2$.

Proof. Any sub-set of W of the form

$$\left\{(x_{t_1},\ldots,x_{t_n}) \in E_1 \times \ldots\ldots \times E_n\right\}, \quad E_i \in \mathcal{B}(S) ,$$

is in $\mathcal{B}(W)$. Since $\mathcal{B}(W)$ is a Borel algebra, any set of the form

$$\left\{(x_{t_1},\ldots,x_{t_n}) \in E^n \in \mathcal{B}(S^n)\right\} .$$

is in $\mathcal{B}(W)$. The class of all sets of the form

$$\left\{(x_{t_1},\ldots,x_{t_n}) \in E^n,\ E^n \in \mathcal{B}(S^n)\right\}$$

for all n , for all n-tuples $0 \leq t_1,\ldots,t_n < \infty$ and all Borel sets E^n of S^n, is an algebra $\mathcal{A}(W) \subset \mathcal{B}(W)$. Further $\mathcal{A}(W)$ generates $\mathcal{B}(W)$.

For fixed $0 \leq t_1,\ldots,t_n < \infty$, let

$$P_a^i(E^n) = P_a^i\left\{(x_{t_1},\ldots,x_{t_n}) \in E^n\right\}, \quad i = 1,2 .$$

Then P_a^i is a measure on the Borel sets of S^n. From (T.6) it follows that $P_a^1(E^n) = P_a^2(E^n)$, for all sets E^n which are finite disjoint unions of sets of the form

$$E_1 \times \ldots\ldots \times E_n , \; E_i \in \mathcal{B}(S).$$

Such sets E^n form an algebra which generates $\mathcal{B}(S^n)$. Using the uniqueness part of the Kolmogoroff theorem, we get $P_a^1(E^n) = P_a^2(E^n)$ for all $E^n \in \mathcal{B}(S^n)$.

Thus $P_a^1 = P_a^2$ on $\mathcal{A}(W)$. One more application of the uniqueness of the extension gives the result.

(T.8) Suppose that $\mathcal{M} = (S, W, P_a, a \in S \cup \{\infty\})$ is a triple with S and W being as in the definition of a Markov process, and P_a, $a \in S \cup \{\infty\}$ are probability distributions on $\mathcal{B}(W)$ and let

$$P(t,a,E) = P_a \{ w : w(t) \in E \}.$$

Suppose further that $p(t,a,E)$ satisfies the properties (T.2), (T.4) and (T.6). Then the contention is that $\mathcal{M}$ is a Markov process with $P(t,a,E)$ as the transition probability of P_a.

To prove this we have to verify conditions b), c) and d) on P_a. The proof of b) is similar to that of (T.7). (T.6) shows that $P_a(B)$ is measurable in a if B is of the form

$$\{(x(t_1),\ldots,x(t_a)) \in E^n,\ E^n \subseteq S^n\}$$

where E^n is a finite disjoint union of sets of the form $E_1 \times E_2 \times \ldots \times E_n$, $E_i \in \mathcal{B}(S)$. For fixed $t_1,\ldots,t_n$, consider the class X of sets $E^n \in \mathcal{B}(S^n)$ for which

$$P_a\{(x_{t_1},\ldots,x_{t_n}) \in E^n\}$$

is measurable in a. If E_i^n is a monotone sequence of sets in X and

$\lim_{i \to \infty} E_i^n = E^n$, $P_a\left\{(x_{t_1},\ldots,x_{t_n}) \in E_i^n\right\}$ is a monotone sequence and

$$\lim_{i \to \infty} P_a\left\{(x_{t_1},\ldots,x_{t_n}) \in E_i^n\right\} = P_a\left\{(x_{t_1},\ldots,x_{t_n}) \in E^n\right\}.$$

X is therefore a monotone class and hence $X \supset \mathcal{B}(S^n)$. We have thus shown that $P_a(B)$ is measurable in a for all $B \in \mathcal{A}(W)$. Similarly we show that the class of sets $B \in \mathcal{B}(W)$ for which $P_a(B)$ is measurable in a, is a monotone class.

We now verify c). Choose $t_n \downarrow 0$ such that

$$P_a\{B_n\} = P_a\left\{x_{t_n} \in U_a\right\} > 1 - \epsilon .$$

Since $w(t)$ is right continuous

$$\left\{w : w(0) \in \overline{U}_a\right\} \supset \bigcap_{n=1}^{\infty} \left[\bigcup_{m=n}^{\infty} B_m \right] .$$

Where $\overline{U}_a$ denotes the closure of U_a. Therefore

$$1 \geqslant P_a(w : w(0) \in \overline{U}_a) \geqslant 1-\epsilon .$$

Since ϵ is arbitrary, $P_a(w : w(0) \in \overline{U}_a) = 1$. Now we choose a decreasing sequence $\left\{U_a^i\right\}$ of open sets such that $U_a^i \supset \overline{U}_a^{i+1}$ and $\bigcap_{i=1}^{\infty} U_a^i = \{a\}$. We then have

$$P_a(x_o = a) = P_a\left(\bigcap_{i=1}^{\infty} (x_o \in U_a^i)\right) = \lim_{i \to \infty} P_a(x_o \in U_a^i) = 1.$$

To prove d) we proceed as follows. First remark that if

$f \in \mathcal{B}(S^n)$ and $E^n \in \mathbb{B}(S^n)$ and $B = ((x_{t_1}, \ldots, x_{t_n}) \in E^n)$ then

$$\int_{E^n} P(t_1,a,da_1)\ldots P(t_n - t_{n-1}, a_{n-1}, da_n) f(a_1,\ldots,a_n) = \int_B f[x_{t_1},\ldots,x_{t_n}]\, d\mathbf{P}_a.$$

Let $B_1 \in \mathcal{B}_t$ be given by $B_1 = (w : w_t^- \in B')$, where

$B' = (x_{t_1'} \in E_1, \ldots, x_{t_n'} \in E_n)$; then $B_1 = (x_{t_i} \in E_i,\ 1 \leq i \leq n)$ with

$t_i = t \wedge t_i'$, $1 \leq i \leq n$. Let $B_2 \in \mathbb{B}$ be given by

$$B_2 = (x_{s_j} \in F_j,\ 1 \leq j \leq m).$$ We have

$$P_a(w \in B_1, w_t^+ \in B_2) = P_a(x_{t_i} \in E_i,\ x_{t+s_j} \in F_j) = P_a(x_{t_i} \in E_i,\ x_t \in S, x_{t+s_j} \in F_j)$$

$$= \int_{\substack{a_i \in E_i \\ c \in S}} P(t_1,a,da_1)\cdots P(t_n - t_{n-1}, a_{n-1}, da_n) P(t - t_n, a_n, dc) \int_{b_j \in F_j} P(s_1,c,db_1)\ldots$$

$$\ldots P(s_m - s_{m-1}, b_{m-1}, db_m)$$

$$= \int_{a_i \in E_i, c \in S} P(t_1,a,da_1)\ldots P(t-t_n,a_n,dc) P_c(B_2) = \int_{B_1} P_{x_t(w)}(B_2) P_a(dw)$$

by the above remark. We now fix B_2 and prove that the above equation holds for all $B_1 \in \mathcal{B}_t$ [the proof runs along the same lines as the proof of b)]. Finally fix $B_1 \in \mathcal{B}_t$ and prove the same for all $B_2 \in \mathbb{B}$.

§4. Semi-groups.

Let $H_t f(a) = \int_S P(t,a,db) f(b) = E_a\{f(x_t)\}$. Then H_t is a map of $\mathcal{B}(S)$ into $\mathcal{B}(S)$ with the following properties:

(H.1) It is linear on $\mathcal{B}(S)$ into $\mathcal{B}(S)$. It is continuous in the sense that if $|f_n| \leq M$ and $f_n \longrightarrow f$ then $H_t f_n \longrightarrow H_t f$.

(H.2) $H_t \geqslant 0$, in the sense that if $f \geqslant 0$, $H_t f \geqslant 0$.

(H.3) It has the semi-group property i.e. $H_t H_s = H_{t+s}$.

$$H_{t+s} f(a) = E_a(f(x_{t+s})) = \int_S P(t+s,a,db)f(b)$$

$$= \int_S f(b) \int_S P(t,a,dc)P(s,c,db)$$

$$= \int_S P(t,a,dc)\left[\int_S f(b)P(s,c,db)\right]$$

$$= \int_S P(t,a,dc)H_s f(c)$$

$$= H_t H_s f(a)$$

(H.4) $H_t 1 \leq 1$

(H.5) $H_t f(a)$ is $\mathcal{B}(R^1)$-measurable in t.

(H.6) If f is continuous at a, $\lim_{t \downarrow 0} H_t f(a) = f(a)$.

For, if U_a is an open set containing a,

$$H_t f(a) = \int_S P(t,a,db)f(b) = \int_{U_a} P(t,a,db)f(a)$$

$$+ \int_{U_a} P(t,a,db)\left[f(b)-f(a)\right] + \int_{S-U_a} P(t,a,db)f(b).$$

$$= f(a)P(t,a,U_a) + \int_{U_a} P(t,a,db)[f(b)-f(a)] + \int_{S-U_a} P(t,a,db)f(b).$$

Now use the fact that $P(t,a,U_a) \longrightarrow 1$ and f is continuous at a.

§ 5. Green operator.

We have seen that the operators $\{H_t\}$ form a semi-group. We now introduce one more operator, the Green operator, as the formal Laplace transform of H_t, which will lead to the concept of a generator.

Consider the operator $G_\alpha = \int_0^\infty e^{-\alpha t} H_t \, dt$, defined for $\alpha > 0$ by

$$G_\alpha f(a) = \int_0^\infty e^{-\alpha t} H_t f(a)dt \, , \quad f \in \mathcal{B}(S).$$

G is called the Green operator on $\mathcal{B}(S)$. Interchanging the orders of integration, we also have

$$G_\alpha f(a) = E_a\left[\int_0^\infty e^{-\alpha t} f(x_t(w))dt\right].$$

Let $G(\alpha, a,E) = \int_0^\infty e^{-\alpha t} P(t,a,E)dt$. This measure on $\mathcal{B}(S)$ is called Green's measure on $\mathcal{B}(S)$. We have

$$G_\alpha f(a) = \int_0^\infty e^{-\alpha t} H_t f(a)dt = \int_S f(b) \int_0^\infty e^{-\alpha t} P(t,a,db)dt$$

$$= \int_S G(\alpha,a,db)f(b).$$

The operator G_α has the following properties:

(G.1) G_α is linear, and continuous in the sense that if $|f_n| < \eta$ and $f_n \longrightarrow f$, then $G_\alpha f_n(a) \longrightarrow G_\alpha f(a)$.

(G.2) $G_\alpha \geq 0$, i.e. $G_\alpha f \geq 0$ if $f \geq 0$.

(G.3) G_α satisfies the following equation, called the <u>resolvent equation</u>:

$$G_\alpha - G_\beta + (\alpha - \beta) G_\alpha G_\beta = 0.$$

We have

$$\begin{aligned} H_s G_\alpha f(a) &= \int_S P(s,a,db) G_\alpha f(b) \\ &= \int_S P(s,a,db) \int_0^\infty e^{-\alpha t} H_t f(b) dt \\ &= \int_0^\infty e^{-\alpha t} H_{t+s} f(a) dt \text{ (interchanging the order of integration)} \\ &= e^{\alpha s} \int_s^\infty e^{-\alpha t} H_t f(a) dt. \end{aligned}$$

Therefore

$$\begin{aligned} G_\beta G_\alpha f(a) &= \int_0^\infty e^{-\beta s} H_s G_\alpha f(a) ds \\ &= \int_0^\infty e^{(\alpha - \beta) s} ds \int_s^\infty e^{-\alpha t} H_t f(a) dt \\ &= \int_0^\infty e^{-\alpha t} H_t f(a) dt \int_0^t e^{(\alpha - \beta) s} ds \\ &= \frac{G_\beta f(a) - G_\alpha f(a)}{\alpha - \beta}. \end{aligned}$$

Remark: $$H_t H_s = H_{t+s} = H_s H_t \quad \text{and}$$

$$G_\alpha G_\beta = \frac{G_\beta - G_\alpha}{\alpha - \beta} = G_\beta G_\alpha$$

(G.4) $G_\alpha 1 \leq \frac{1}{\alpha}$, because $H_t 1 \leq 1$

(G.5) The integral defining G_α exists for complex numbers α whose real part > 0 for every $f \in \mathcal{B}(S)$. Then $G_\alpha f(a)$ is analytic in α for every $f \in \mathcal{B}(S)$ and every $a \in S$.

(G.6) f is continuous at a implies

$$\alpha G_\alpha f(a) \longrightarrow f(a) \quad \text{as } \alpha \longrightarrow \infty \ .$$

For $\alpha G_\alpha f(a) = \int_0^\infty \alpha e^{-\alpha t} H_t f(a) dt = \int_0^\infty e^{-t} H_{t/\alpha} f(a) dt$

and $H_t f(a) \longrightarrow f(a)$ as $t \longrightarrow 0$ if f is continuous at a.

§ 6. The Generator.

Define, for $f \in \mathcal{B}(S)$

$$\| f \| = \sup_{a \in S} | f(a) | \ .$$

Then $$| H_t f(a) | \leq \| f \| \ .$$

$\mathcal{B}(S)$ is a Banach space with the norm $\| f \|$, and H_t becomes a semi-group of continuous linear operators on $\mathcal{B}(S)$.

Consider the following purely formal calculations.

$$\mathfrak{G} = \lim_{t \to 0} \frac{H_t - I}{t} = \left[\frac{dH_t}{dt} \right]_{t=0}$$

Then $$\frac{dH_t}{dt} = \lim_{\delta \to 0} \frac{H_{t+\delta} - H_t}{\delta} = \lim_{\delta \to 0} \frac{H_\delta - I}{\delta} . H_t = \mathcal{G} H_t .$$

Therefore $H_t = e^{t\mathcal{G}}$ and

$$G_\alpha = \int_0^\infty e^{-\alpha t} H_t dt = \int_0^\infty e^{-(\alpha - \mathcal{G})t} dt = (\alpha - \mathcal{G})^{-1}$$

or $$\mathcal{G} = \alpha - G_\alpha^{-1} .$$

The above purely formal calculations have been given precise meaning, and the steps justified by Hille and Yosida **whenever** H_t satisfy certain conditions. In our case, however, H_t do not in general satisfy these conditions, and we shall define $\mathcal{G}$ with the last equation in view. We now proceed to the rigorous definition.

Let $\mathcal{R}_\alpha = G_\alpha[\mathcal{B}(S)]$, $\mathcal{N}_\alpha = G_\alpha^{-1}\{0\}$ be the image and kernel of G_α respectively. We show that $\mathcal{R}_\alpha$ and $\mathcal{N}_\alpha$ are independent of α and that $\mathcal{R}_\alpha \cap \mathcal{N}_\alpha = \{0\}$. The resolvent equation gives

$$G_\alpha f - G_\beta f + (\alpha - \beta) G_\alpha G_\beta f = 0$$

$$\text{i.e. } G_\beta f = G_\alpha [f + (\alpha - \beta) G_\beta f] .$$

Since $f + (\alpha - \beta) G_\beta f \in \mathcal{B}(S)$, it follows that

$$G_\beta f \in G_\alpha [\mathcal{B}(S)] = \mathcal{R}_\alpha ,$$

or that $\mathcal{R}_\beta \subset \mathcal{R}_\alpha$. Interchanging the roles of α and β, $\mathcal{R}_\beta \supset \mathcal{R}_\alpha$ or $\mathcal{R}_\alpha \equiv \mathcal{R}_\beta$. We denote $G_\alpha[\mathcal{B}(S)]$ by $\mathcal{R}$. Similarly

$f \in \mathcal{N}_\beta$ gives $G_\beta f = 0$ and the resolvent equation then gives $G_\alpha f = 0$ or $\mathcal{N}_\beta \subset \mathcal{N}_\alpha$. We denote $G_\alpha^{-1}\{0\}$ by $\mathcal{N}$. Let $u \in \mathcal{R} \cap \mathcal{N}$. Then $u = G_\alpha f$ for some $f \in \mathcal{B}(S)$, and for every β, $G_\beta u = 0$. Now

$$H_s u(a) = H_s G_\alpha f(a) = e^{\alpha s} \int_s^\infty e^{-\alpha t} H_t f(a) dt ,$$

and so $H_s u(a)$ is continuous in s and $\longrightarrow u(a)$ as $s \longrightarrow 0$. Also, since $\int_0^\infty e^{-\beta s} H_s u(a) ds = G_\beta u(a) = 0$ for all β, $H_s u(a) \equiv 0$. Letting $s \longrightarrow 0$ we see that $u(a) \equiv 0$.

For $u \in \mathcal{R}$, define

$$\mathcal{G}_\alpha u = \alpha u - G_\alpha^{-1} u .$$

$\mathcal{G}_\alpha u$ is then determined mod $\mathcal{N}$. We now prove that $\mathcal{G}_\alpha u$ is independent of α. If $f = \mathcal{G}_\alpha u \pmod{\mathcal{N}}$ then $f = \alpha u - G_\alpha^{-1} u \cdot$ (mod $\mathcal{N}$) and

$$G_\alpha f = \alpha G_\alpha u - u ,$$

$$G_\beta G_\alpha f = \alpha G_\beta G_\alpha u - G_\beta u ,$$

$$\frac{G_\alpha - G_\beta}{\beta - \alpha} f = \alpha \frac{G_\alpha - G_\beta}{\beta - \alpha} u - G_\beta u ,$$

$$G_\alpha f - G_\beta f = \alpha G_\alpha u - \beta G_\beta u ,$$

$$G_\beta f = G_\alpha f - \alpha G_\alpha u + \beta G_\beta u$$

$$= - u + \beta G_\beta u$$

$$f = \beta u - G_\beta^{-1} u \pmod{\mathcal{N}} = \mathcal{G}_\beta u \pmod{\mathcal{N}}.$$

We denote $\mathcal{G}_\alpha u$ by $\mathcal{G} u$. Then if $G_\alpha f = u$ we have

$$\mathcal{G} u = \alpha u - f \pmod{\mathcal{N}}.$$

Thus $u = G_\alpha f$ if and only if $(\alpha - \mathcal{G}) u = f \pmod{\mathcal{N}}$. The domain $\mathcal{D}(\mathcal{G})$ of $\mathcal{G}$ is $\mathcal{R}$ and we have $\mathcal{G} = \alpha - G_\alpha^{-1}$. $\mathcal{G}$ is called the <u>generator</u> of the Markov process.

The following theorem shows that the generator determines the Markov process uniquely.

<u>Theorem</u>

Let $\mathcal{M}_i = (S, W, P_a^i, a \in S \cup \{\infty\})$, $i = 1,2$, be two Markov processes, and $\mathcal{G}_i$, $i = 1,2$ their generators. Then if $\mathcal{G}_1 = \mathcal{G}_2$, $P_a^1 = P_a^2$, i.e. $\mathcal{M}_1 = \mathcal{M}_2$.

<u>Proof</u>. $\mathcal{D}(\mathcal{G}_i) = G_\alpha^i[\mathcal{B}(S)] = \mathcal{R}^i$. Since $\mathcal{G}_1 = \mathcal{G}_2$, $\mathcal{D}(\mathcal{G}_1) = \mathcal{D}(\mathcal{G}_2)$, i.e. $\mathcal{R}^1 = \mathcal{R}^2 = \mathcal{R}$ (say). Since their ranges must also be the same, $\mathcal{N}_1 = \mathcal{N}_2 = \mathcal{N}$ (say). We have therefore

$$(\alpha - \mathcal{G}_1) G_\alpha^1 f = f \pmod{\mathcal{N}}$$

$$(\alpha - \mathcal{G}_2) G_\alpha^2 f = f \pmod{\mathcal{N}}$$

$$(\alpha - \mathcal{G}_1) G_\alpha^1 f = (\alpha - \mathcal{G}_2) G_\alpha^2 f \pmod{\mathcal{N}},$$

$$(\alpha - \mathcal{G}) G_\alpha^1 f = (\alpha - \mathcal{G}) G_\alpha^2 f \pmod{\mathcal{N}} \text{ since } \mathcal{G}_1 = \mathcal{G}_2$$

By definition $\alpha - \mathcal{G} = \alpha - \mathcal{G}_1 = G_\alpha^{1^{-1}}$. Therefore

$$G_\alpha^{1^{-1}} G_\alpha^1 f = G_\alpha^{1^{-1}} G_\alpha^2 f \qquad (\text{mod } \mathcal{N})$$

$$G_\alpha^1 G_\alpha^{1^{-1}} G_\alpha^1 f = G_\alpha^1 G_\alpha^{1^{-1}} G_\alpha^2 f .$$

Therefore $G_\alpha^1 f = G_\alpha^2 f$. This gives

$$\int_0^\infty e^{-\alpha t} H_t^1 f(a)dt = \int_0^\infty e^{-\alpha t} H_t^2 f(a)dt \quad \text{for every} \quad f \in \mathcal{B}(S).$$

Thus if f is continuous, $H_t^1 f(a) \equiv H_t^2 f(a)$

$$\int P^1(t,a,db)f(b) = \int P^2(t,a,db)f(b)$$

for every $f \in \mathcal{B}(S)$ which is continuous. Therefore

$$P^1(t,a,E) = P^2(t,a,E) ,$$

Hence

$$P_a^1 = P_a^2 .$$

§7. Examples.

We first prove a lemma which will have applications later, and then we give a few examples of Markov processes.

Let f be a real-valued function on an open interval (a,b). When f is of bounded variation in every compact sub-interval of (a,b) we write $f \in \mathcal{B}W(a,b)$ and then there exists a unique signed measure df (Lebesgue-Stieltjes measure) such that $df(\alpha, \beta] = f(\beta+) - f(\alpha+)$, $(\alpha, \beta] \subseteq (a,b)$. Suppose that μ is any measure on (a,b) which is finite on compact subsets of (a,b). Suppose further that there exists a function φ on (a,b) which is μ-summable on every compact sub-interval of (a,b) and satisfies

$$f(\beta+) - f(\alpha+) = \int_{\alpha}^{\beta} \varphi(\xi) d\mu(\xi).$$

Then $df = \varphi\, d\mu$ and f is absolutely continuous with respect to $d\mu$.

We now prove the following

<u>Lemma</u>. If $f,g \in \mathcal{B}W(a,b)$ then $fg \in \mathcal{B}W(a,b)$ and

$$d(fg)x = f(x+)dg(x) + g(x-)df(x).$$

<u>Proof</u>. We can assume that f and g are non-negative and non-decreasing in (a,b). It is enough to prove that if h is continuous in (a,b) and has compact support, then,

$$\int h(x)d(fg)(x) = \int h(x)f(x+)dg(x) + \int h(x)g(x-)df(x).$$

For n = 1,2,.... let $\{\alpha_{n,k}\}$, k = 0, $\pm$ 1, $\pm$ 2,.... be a sequence of points such that

$$a \leftarrow \dots < \alpha_{n,o} < \alpha_{n,1} < \dots \rightarrow b\, , \; \alpha_{n,i} - \alpha_{n,i-1} < \frac{1}{n}$$

Define

$$\begin{aligned} \varphi_n(x) &= \alpha_{n,i} \\ \psi_n(x) &= \alpha_{n,i-1} \end{aligned} \qquad \text{if } \alpha_{n,i-1} < x < \alpha_{n,i}.$$

Then

$$I_n = \int h[\varphi_n(x)]\, d(f.g)(x)$$

$$= \sum_{i=-\infty}^{\infty} h(\alpha_{n,i}) \left[f(\alpha_{n,i}+)g(\alpha_{n,i}+)-f(\alpha_{n,i-1}+)g(\alpha_{n,i-1}+)\right]$$

$$= \sum_{i=-\infty}^{\infty} h(\alpha_{n,i})f(\alpha_{n,i}+)\left[g(\alpha_{n,i}+)-g(\alpha_{n,i-1}+)\right] +$$

$$+ \sum_{i=-\infty}^{\infty} h(\alpha_{n,i})\, g(\alpha_{n,i-1}+)\left[f(\alpha_{n,i}+)-f(\alpha_{n,i-1}+)\right]$$

$$= \int h[\varphi_n(x)]\, f[\varphi_n(x)+]\, dg(x)$$

$$+ \int h[\psi_n(x)]\, g[\psi_n(x)+]\, df(x)$$

Since h has compact support, letting $n \longrightarrow \infty$ we get the result.

Ex.1 Standard Brownian motion

Let $S = R^1$, $W = C[0,\infty)$ [we define $w(\infty) = \infty$] . Let P be the Wiener measure on W and define for $a \in S$,

$$\mathbf{P}_a(B) = P\{w: w+a \in B\}, \quad B \in \mathcal{B}(W).$$

It is not difficult to show that (S, W, P_a) is a Markov process; this is a continuous process and is called the Standard Brownian motion.

We shall determine the generator of this process. We have

$$P(t,a,E) = P\{w:\ w_t + a \in E\}\ \frac{1}{\sqrt{2\pi t}} \int_{E-a} e^{-x^2/2t}\, dx$$

$$= \int_E N(t,a,c)\, dc\ .$$

$$H_t f(a) = \int_{R^1} N(t,a,b) f(b)db = \int_{-\infty}^{\infty} \frac{e^{-(b-a)^2/2t}}{\sqrt{2\pi t}}\ f(b)db.$$

If $u \in \mathcal{R}$, $u = G_\alpha f$ for some $f \in \mathcal{B}(R^1)$ and

$$u(a) = G_\alpha f(a) = \int_0^\infty e^{-\alpha t} H_t f(a)dt = \int_{-\infty}^{\infty} f(b)db \int_0^\infty \frac{e^{-\alpha t - \frac{(b-a)^2}{2t}}}{\sqrt{2\pi t}}\ dt$$

$$= \int_{-\infty}^{\infty} \frac{1}{\sqrt{2\alpha}}\ e^{-\sqrt{2\alpha}\,|b-a|}\ f(b)db$$

$$= e^{-\sqrt{2\alpha}\,a} \int_{-\infty}^{a} \frac{1}{\sqrt{2\alpha}}\ e^{\sqrt{2\alpha}\,b} f(b)db + e^{\sqrt{2\alpha}\,a} \int_a^\infty \frac{1}{\sqrt{2\alpha}}\ e^{-\sqrt{2\alpha}\,b} f(b)db.$$

Since $e^{-\sqrt{2\alpha}\,a}$ and $\int_{-\infty}^{a} e^{\sqrt{2\alpha}\,b} f(b)db$ are both in $\mathcal{B}W(-\infty,\infty)$ we get from the lemma

$$du(a) = -\sqrt{2\alpha}\, e^{-\sqrt{2\alpha}\,a} da \int_{-\infty}^{a} \frac{1}{\sqrt{2\alpha}}\ e^{\sqrt{2\alpha}\,b} f(b)db + \frac{e^{-\sqrt{2\alpha}a}\ e^{\sqrt{2\alpha}\,a}\ f(a)}{\sqrt{2\alpha}}\ da$$

$$+ \sqrt{2\alpha}\, e^{\sqrt{2\alpha}\,a} da \int_a^\infty \frac{1}{\sqrt{2\alpha}}\ e^{-\sqrt{2\alpha}\,b} f(b)db - \frac{e^{\sqrt{2\alpha}\,a}\ e^{-\sqrt{2\alpha}\,a}\ f(a)}{\sqrt{2\alpha}}\ da.$$

Therefore u is absolutely continuous and

$$u'(a) = -e^{-\sqrt{2\alpha}\,a}\int_{-\infty}^{a} e^{\sqrt{2\alpha}\,b} f(b)\,db + e^{\sqrt{2\alpha}\,a}\int_{a}^{\infty} e^{-\sqrt{2\alpha}\,b} f(b)\,db$$

almost everywhere. Using the lemma again we see that u' is absolutely continuous and we get

$$u'' = 2\alpha u - 2f \quad \text{almost everywhere.}$$

Let $\mathcal{R}_+ = \{u : u \in \mathcal{B}(R^1),\ u \text{ abs. cont, } u' \text{ abs.cont, } u'' \in \mathcal{B}(R^1)\}$
We have seen above that if $u \in \mathcal{R}$, then $u \in \mathcal{R}_+$. Conversely let $u \in \mathcal{R}_+$ and put $f = \alpha u - \frac{1}{2} u''$. Then $f \in \mathcal{B}(R^1)$ and $v = G_\alpha f$ satisfies

$$\tfrac{1}{2} v'' = \alpha v - f$$

Therefore $w = v-u$ satisfies

$$\tfrac{1}{2} w'' - \alpha w = 0.$$

Hence $w = c_1 e^{\sqrt{2\alpha}\,a} + c_2 e^{-\sqrt{2\alpha}\,a}$. Since w is bounded, $c_1 = c_2 = 0$ or $u = G_\alpha f$. Thus we have proved that

$$\mathcal{R} = \{u : u \in \mathcal{B}(R^1),\ u \text{ abs.cont, } u' \text{ abs.cont, } u'' \in \mathcal{B}(R^1)\}$$

If $f \in \mathcal{N}$, $u = G_\alpha f = 0$ and since $u'' = 2\alpha u - 2f$ a.e. we see that $f = 0$ a.e. Therefore

$$\mathcal{N} = \{f : f = 0 \ \text{ a.e.}\}.$$

Also the formula $u'' = 2\alpha u - 2f$ (a.e.) shows that $\mathcal{G} u = \frac{u''}{2}$ (a.e.) and hence $\mathcal{G} = \frac{1}{2} \frac{d^2}{da^2}$.

Ex. 2 Brownian motion with reflecting barrier at t = 0.

Let $(\hat{S} = (-\infty, \infty), \hat{W}, \hat{P}_a)$ denote the Standard Brownian motion.

Let $S = [0,\infty)$ and W the set of all continuous functions on $[0,\infty)$ into S. If $B \in \mathcal{B}(W)$ then $B \in \mathcal{B}(\hat{W})$. Define $P_a(B) = \hat{P}_a [w : |w| \in B]$ for $a \in S$. Then (S,W,P_a) is a continuous Markov process and is called the Brownian motion with reflecting barrier at t = 0.

We have

$$P(t,a,E) = \hat{P}_a \left\{ w : |w(t)| \in E \right\}$$

$$= \hat{P}_a \left\{ w : w(t) \in E \cup (-E) \right\}$$

$$= \int_E [N(t,a,b) + N(t,a,-b)] \, db$$

$$H_t f(a) = \int_0^\infty \left[N(t,a,b) + N(t,a,-b)\right] f(b)db$$

$$= \int_{-\infty}^{\infty} N(t,a,b) \hat{f}(b)db = \hat{H}_t \hat{f}(a)$$

where $\hat{f}(b) = f(|b|)$. Therefore.

$$u(a) = G_\alpha f(a) = \int_0^\infty e^{-\alpha t} H_t f(a)dt = \int_0^\infty e^{-\alpha t} \hat{H}_t \hat{f}(a)dt = \hat{G}_\alpha \hat{f}(a) = \hat{u}(a), \text{ say.}$$

From the previous example it follows that $\hat{u} \in \mathcal{B}(\hat{S})$, $\hat{u}$ is absolutely continuous, $\hat{u}'$ is absolutoly continuous and $\hat{u}'' \in \mathcal{B}(\hat{S})$. Since $u(a) = \hat{u}(a)$ for $a > 0$, we see that $u \in \mathcal{B}(S)$, u is absolutely continuous for $a > 0$, u' is absolutely continuous, $u'' \in \mathcal{B}(S)$. Further since $\hat{u}(a) = \hat{u}(-a)$ we see that $\hat{u}'(a) = -\hat{u}'(-a)$ and hence $\hat{u}'(0) = 0$. This gives $u^{+}(0) = 0$. The relation $\frac{1}{2}\hat{u}'' = \alpha\hat{u} - \hat{f}$ gives $\frac{1}{2}u'' = \alpha u - f$, where $u^{+}(0) = \lim\limits_{h \downarrow 0} \dfrac{u(h) - u(0)}{h}$.

$$\mathcal{N} = \left\{ f : f = 0 \text{ a.e.} \right\}.$$

$$\mathcal{R} = \left\{ u : u \in \mathcal{B}(S),\ u,\ u' \text{ abs.cont},\ u^{+}(0) = 0 \text{ and } u'' \in \mathcal{B}(S) \right\}.$$

$$\mathcal{G}u = \alpha u - f = \tfrac{1}{2}u'' \quad (\text{a.e.})$$

<u>Ex. 3</u> <u>Poisson process</u>

Let (Ω, P) be a probability measure space and $\left\{ \xi(t,\omega),\ 0 \leq t < \infty \right\}$ a Poisson process on Ω.

Let $S = \left\{0,1,2,\ldots\ldots\right\}$, $W = W_{d_1}$ = the set of all sample paths whose only discontinuities are of the first kind, and hence they are step functions with integral values. For almost all ω, $\xi(t,\omega)$ is a step function with jump 1 and vanishes at $t = 0$; therefore, for almost all ω, $\xi(t,\omega)$ is a step function with integral values and hence belongs to W.

Let $\eta^{(k)}(t,\omega) = k + \xi(t,\omega)$ and define

$$P_k(B) = P\left\{\omega : \eta^{(k)}(\cdot,\omega) \in B\right\}, \quad B \in \mathcal{B}(W).$$

If $E \subset S$,

$$P(t,k,E) = P(\omega : k + \xi(t,\omega) \in E)$$

$$= P(\omega : \xi(t,\omega) \in E - k)$$

$$= \sum_{0 \leq n \in E-k} e^{-\lambda t} \frac{(\lambda t)^n}{n!}$$

$$= \sum_{k \leq n \in E} e^{-\lambda t} \frac{(\lambda t)^{n-k}}{(n-k)!}$$

$$H_t f(k) = \sum_{n=0}^{\infty} f(n+k) e^{-\lambda t} \frac{(\lambda t)^n}{n!}$$

$$u(k) = G_\alpha f(k) = \int_0^\infty e^{-\alpha t} \sum f(n+k) e^{-\lambda t} \frac{(\lambda t)^n}{n!} dt$$

$$= \sum_{n=0}^{\infty} f(n+k) \frac{\lambda^n}{(\alpha+\lambda)^{n+1}} .$$

Therefore we obtain

$$u(k+1) - u(k) = - \frac{f(k)}{\lambda} + \frac{\alpha}{\lambda} u(k).$$

If $u = G_\alpha f = 0$, from the above we see that $f \equiv 0$, and

$$\mathfrak{N} = \{f : f \equiv 0\}.$$

Let $u \in \mathfrak{B}(S)$ and put $f(k) = \alpha u(k) - \lambda [u(k+1)-u(k)]$. If $v(k) = G_\alpha f(k)$, v satisfies

$$v(k+1) - v(k) = - \frac{f(k)}{\lambda} + \frac{\alpha}{\lambda} v(k)$$

and hence, subtracting,

$$\alpha\left[v(k)-u(k)\right] - \lambda\left[v(k+1)-u(k+1)-v(k)+u(k)\right] = 0$$

and so

$$v(k+1)-u(k+1) = \frac{\alpha+\lambda}{\lambda}\left[v(k)-u(k)\right] .$$

If $v(0) \neq u(0)$, $|v(k)-u(k)| = \left(\frac{\alpha+\lambda}{\lambda}\right)^k |v(0)-u(0)| \longrightarrow \infty$ which is impossible since $v-u \in \mathcal{B}(S)$. Therefore $v(0) = u(0)$ and hence $v(k) = u(k)$. Thus we have $\mathcal{R} = \mathcal{B}(S)$.

Ex. 4 Constant velocity motion

Let $S = R^1$, $W = C\,[0, \infty)$. Let

$$P_a\left\{ w(t) \equiv a+\lambda t, \ 0 \leq t < \infty \right\} = 1.$$

Then for any $B \in \mathcal{B}$ if $w(t) = a+\lambda t \in B$, $P_a(B) = 1$ and otherwise $P_a(B) = 0$.

$$P(t,a,E) = \delta(E, a+\lambda t) = \begin{cases} 1 & \text{if } a+\lambda t \in E \\ 0 & \text{if } a+\lambda t \notin E \end{cases}$$

$$H_t f(a) = f(a+\lambda t)$$

$$u(a) = G_\alpha f(a) = \frac{1}{\lambda} e^{\frac{\alpha}{\lambda} a} \int_a^\infty e^{-\frac{\alpha}{\lambda} t} f(t)\,dt.$$

From the lemma and the absolute continuity of u ,

$$u'(a) = \frac{\alpha}{\lambda} u(a) - \frac{f(a)}{\lambda} \quad (\text{a.e.}).$$

So if $G_\alpha f = 0$, $f = 0$ a.e.

$$\mathcal{N} = \{ f : f = 0 \quad \text{a.e.} \}$$

$$\mathcal{R} \subset \{ u : u \in \mathcal{B}(R^1), u \text{ abs.cont}, \; u' \in \mathcal{B}(R^1) \}$$

$$\mathcal{G} u = \alpha u - f = \lambda u',$$

so that $\mathcal{G} = \lambda \frac{d}{da}$.

If $u \in \mathcal{R}$, we have $u \in \mathcal{B}(R^1)$, u abs.cont. and $u' \in \mathcal{B}(R^1)$. Conversely, let u satisfy these conditions and $f = \lambda u - u'$. Then $v = G_\alpha f$ satisfies

$$\alpha y - \lambda y = f .$$

The general solution therefore is

$$y = G_\alpha f + C e^{\frac{\alpha}{\lambda} a} .$$

Since y is to be bounded, $C = 0$. Thus

$$\mathcal{R} = \{ u : u \in \mathcal{B}(R^1), \; u \text{ abs.cont}, \; u' \in \mathcal{B}(R^1) \} .$$

Ex. 5 Positive velocity motion

Let $S = (r_1, r_2)$ and $v(x) > 0$ a function continuous on (r_1, r_2) such that for $r_1 < \alpha < \beta < r_2$

$$\int_{\alpha}^{\beta} \frac{dx}{v(x)} < +\infty \quad \text{and} \quad \int^{r_2} \frac{dx}{v(x)} = +\infty .$$

Then there exists a solution $\xi^{(a)}(t)$ of $\frac{d\xi}{dt} = v(\xi)$ with the initial condition $\xi^{(a)}(0) = a$.

Let $W = W_c$ and

$$P_a\{x_t(w) \equiv \xi^{(a)}(t), \quad 0 \le t < \infty\} = 1.$$

This is similar to Ex.4 and we can proceed on the same lines.

§8. Dual notions.

Let $\mathbb{M} = (S,W,P_a)$ be a Markov process and $\mathfrak{M}$ the set of all bounded signed measures on $\mathbb{B}(S)$. $\mathfrak{M}$ is a linear space. For $E \in \mathbb{B}(S)$ and $\mu \in \mathfrak{M}$, define

$$\|\mu\| = \text{total variation of } \mu = \sup_{E \in \mathbb{B}(S)} [\mu(E) - \mu(E^c)] .$$

$$H_t^* \mu(E) = \int_S P(t,a,E)\, \mu(da)$$

$$G_\alpha^* \mu(E) = \int_0^\infty e^{-\alpha t} H_t^* \mu(E) dt.$$

Then $H_t^* \mu$ and $G_\alpha^* \mu$ are in $\mathfrak{M}$ and

$$\| H_t^* \mu \| \le \|\mu\|, \; \| G_\alpha^* \mu \| \le \frac{\|\mu\|}{\alpha}$$

Also, for $f \in \mathfrak{B}(S)$, denote by $(f, H_t^* \mu)$ and $(f, G_\alpha^* \mu)$ the integrals $\int f(a) H_t^* \mu(da)$ and $\int f(a) G_\alpha^* \mu(da)$ respectively. We have

$$(f,H_t^*\mu) = \int f(a)H_t^*\mu(da) = \iint f(a)P(t,b,da)\,\mu(db)$$

$$= \int H_t f(b)\,\mu(db) = (H_t f,\mu).$$

Similarly $\quad (f,G_\alpha^*\mu) = (G_\alpha f,\mu).$

Theorem 1.

$$G_\alpha^* - G_\beta^* + (\alpha-\beta)\,G_\alpha^*\,G_\beta^* = 0.$$

Follows easily from $(f,G_\alpha^*\mu) = (G_\alpha f,\mu)$ and the resolvent equation for G_α .

Theorem 2

$\mathcal{R}_\alpha^* = G_\alpha^*\mathcal{M}$ is independent of α . We denote this by $\mathcal{R}^*$.

Follows from Theorem 1.

Theorem 3

If $G_\alpha^*\mu = 0$, $\mu \in \mathcal{M}$, then $\mu = 0$.

Proof. Let $f \in C(S)$. Then since $\alpha\, G_\alpha^*\mu = 0$ we have

$$0 = (f,\alpha\, G_\alpha^*\mu) = (\alpha G_\alpha f,\mu) \longrightarrow (f,\mu)$$

as $\alpha \longrightarrow \infty$. Hence, for every $f \in C(S)$, $(f,\mu) = 0$. It follows that $\mu \equiv 0$.

Theorem 4

$\mathcal{G}_\alpha^* = \alpha - (G_\alpha^*)^{-1}$ is independent of α . We denote this by $\mathcal{G}^*$, and call it the dual generator of $\mathcal{G}$.

Proof is easy.

<u>Theorem 5</u>

If $u \in \mathcal{R} = \mathcal{D}(\mathcal{G}), \nu \in \mathcal{R}^* = \mathcal{D}(\mathcal{G}^*)$ then

$$(\mathcal{G} u, \nu) = (u, \mathcal{G}^* \nu).$$

<u>Proof</u>. Let $u = G_\alpha f$, $\nu = G_\alpha^* \mu$. Then

$$(\mathcal{G} u, \nu) = (\alpha u - f, \nu) = (\alpha u, \nu) - (f, \nu)$$

$$= (\alpha u, \nu) - (f, G_\alpha^* \mu) = (u, \alpha \nu) - (G_\alpha f, \mu)$$

$$= (u, \alpha \nu) - (u, \mu) = (u, \alpha \nu - \mu) = (u, \mathcal{G}^* \nu).$$

This theorem justifies the name <u>dual generator</u> for $\mathcal{G}^*$.

<u>Theorem 6</u>

$\mathcal{G}^*$ determines the Markov process i.e. if $\mathcal{M}_i = (S^i, W^i, P_a^i), i=1,2$ are two Markov processes with $S^1 = S^2$, $W^1 = W^2$ and $\mathcal{G}^{1*} = \mathcal{G}^{2*}$, then $P_a^1 = P_a^2$.

<u>Proof</u>. Since $\mathcal{G}^{1*} = \mathcal{G}^{2*}$, $\mathcal{D}(\mathcal{G}^{1*}) = \mathcal{D}(\mathcal{G}^{2*})$. Let $\mu \in \mathcal{M}$ and $\nu = G_\alpha^{1*} \mu$. Since $\nu \in \mathcal{D}(\mathcal{G}^{2*})$, $\nu = G_\alpha^{2*} \mu_1$. Now $\alpha\nu - \mu = \mathcal{G}^{1*}\nu = \mathcal{G}^{2*}\nu = \alpha\nu - \mu_1$. Hence $\mu = \mu_1$ i.e. $G_\alpha^{1*}\mu = G_\alpha^{2*}\mu$. Now for any $f \in \mathcal{B}(S)$, and for any $\mu \in \mathcal{M}$,

$$(G_\alpha^1 f, \mu) = (f, G_\alpha^{1*}\mu) = (f, G_\alpha^{2*}\mu) = (G_\alpha^2 f, \mu).$$

It follows that $G_\alpha^1 f \equiv G_\alpha^2 f$, i.e. $H_t^1 f(a) = H_t^2 f(a)$ for almost all t. If $f \in C(S)$, $H_t^i f(a)$ $i=1,2$ are right continuous in t and are equal almost everywhere. They are therefore identical. Now the proof can be completed easily.

Example. Consider the standard Brownian motion. Then

$$\mathcal{R}^{*} = \left\{ \nu : \nu(db) = db \int \mu(da)\, G(\alpha, |a-b|) \right\}.$$

This means $\nu(E) = \int_E db \int \mu(da)\, G(\alpha, |a-b|)$ where

$$G(\alpha, |a-b|) = \int_0^{\infty} e^{-\alpha t} N(t,a,b)\,dt = \frac{1}{\sqrt{2\alpha}}\, e^{-\sqrt{2\alpha}|b-a|}.$$

The formula shows that ν has the density

$$u(b) = \int_{-\infty}^{\infty} \frac{1}{\sqrt{2\alpha}}\, e^{-\sqrt{2\alpha}|b-a|}\, \mu(da)$$

$$= e^{-\sqrt{2\alpha}\, b} \int_{-\infty}^{b} \frac{1}{\sqrt{2\alpha}}\, e^{\sqrt{2\alpha}\, a}\, \mu(da) + e^{\sqrt{2\alpha}\, b} \int_{b}^{\infty} \frac{1}{\sqrt{2\alpha}}\, e^{-\sqrt{2\alpha}\, a}\, \mu(da).$$

Now using the lemma of §7, we see that

$$du(b) = -e^{-\sqrt{2\alpha}\, b}\, db \int_{-\infty}^{b} e^{\sqrt{2\alpha}\, a}\, \mu(da) + e^{\sqrt{2\alpha}\, b}\, db \int_{b}^{\infty} e^{-\sqrt{2\alpha}\, a}\, \mu(da)$$

and hence u is absolutely continuous and

$$u'(b) = -e^{-\sqrt{2\alpha}\, b} \int_{-\infty}^{b} e^{\sqrt{2\alpha}\, a}\, \mu(da) + e^{\sqrt{2\alpha}\, b} \int_{b}^{\infty} e^{-\sqrt{2\alpha}\, a}\, \mu(da)$$

Using the same lemma again we see that

$$du'(b) = -2\mu(db) + \sqrt{2\alpha}\, db \int_{-\infty}^{\infty} e^{-\sqrt{2\alpha}|b-a|}\, (da)$$

$$= -2\mu(db) + 2\alpha\nu(db).$$

Thus we have $\mathcal{G}^*\nu = \alpha\nu - \mu = \frac{1}{2}\,du'$.

§9. A Theorem of Kac.

We prove the following

Theorem (Kac).

Let $M = (S,W,P_a)$ be a Markov process. For $k,f \in \mathcal{B}(S)$ we define

$$v(a) = v(\alpha,a) = E_a\left[\int_0^\infty e^{-\alpha t} f(x_t) e^{-\int_0^t k(x_s)ds}\,dt\right]$$

where $\alpha > \|k^-\| = \sup(-k(a) \vee 0)$, $\{(a \vee b) := \max(a,b)\}$. Then

$$(k + \alpha - \mathcal{G})v = f .$$

[If $k \geqslant 0$, $\|k^-\| = 0$ and $\alpha > 0$].

Proof. Let $u(\alpha,a) = G_\alpha f(a)$. Then we have

$$v-u = E_a\left(\int_0^\infty e^{-\alpha t} f(x_t)\left[e^{-\int_0^t k(x_s)ds} - 1\right]dt\right)$$

$$= -E_a\left(\int_0^\infty e^{-\alpha t} f(x_t)\left[\int_0^t e^{-\int_s^t k(x_\theta)d\theta} k(x_s)ds\right]dt\right)$$

Now

$$\int_0^\infty \int_0^t \left| e^{-\alpha t} f(x_t)\, e^{-\int_s^t k(x_\theta)d\theta} k(x_s)\right| ds\,dt$$

$$\leq \int_0^\infty \|f\|\,\|k\|\, e^{-(\alpha - \|k^-\|)t}\, t\,dt < \infty .$$

Changing the order of integration

$$v-u = -E_a\left(\int_0^\infty k(x_s)\,ds \int_s^\infty e^{-\alpha t} f(x_t)\, e^{-\int_s^t k(x_\theta)\,d\theta}\,dt\right)$$

$$= -E_a\left(\int_0^\infty k(x_s)\,ds \int_0^\infty e^{-\alpha(t+s)} f(x_{t+s})\, e^{-\int_s^{t+s} k(x_\theta)\,d\theta}\,dt\right)$$

$$= -E_a\left(\int_0^\infty e^{-\alpha s} k(x_s)\,ds \int_0^\infty e^{-\alpha t} f(x_{t+s})\, e^{-\int_0^{t} k(x_{\theta+s})\,d\theta}\,dt\right)$$

$$= -E_a\left(\int_0^\infty e^{-\alpha s} k(x_s)\,ds \int_0^\infty e^{-\alpha t} f[x_t(w_s^+)]\, e^{-\int_0^{t} k[x_\theta(w_s^+)]\,d\theta}\,dt\right)$$

$$= -\int_0^\infty e^{-\alpha s}\,ds E_a\left[k(x_s) \int_0^\infty e^{-\alpha t} f[x_t(w_s^+)]\, e^{-\int_0^{t} k[x_\theta(w_s^+)]\,d\theta}\,dt\right]$$

$$= -\int_0^\infty e^{-\alpha s}\,ds E_a\left[k(x_s) E_{x_s}\left(\int_0^\infty e^{-\alpha t} f(x_t)\, e^{-\int_0^{t} k(x_\theta)\,d\theta}\,dt\right)\right]$$

$$= -\int_0^\infty e^{-\alpha s}\,ds\, E_a\left[\, k(x_s) v(x_s)\right]$$

$$= -\, G_\alpha(k\, v)(a) \in \mathcal{D}(\mathcal{G}).$$

Since $u \in \mathcal{D}(\mathcal{G})$,

$v \in \mathcal{D}(\mathcal{G})$.

Further

$$G_{\alpha}^{-1}[v-u] = -kv \ (\text{mod } (\mathcal{N})) \quad \text{and} \quad G_{\alpha}^{-1} = \alpha - \mathcal{G}$$

$$(\alpha - \mathcal{G})(v-u) = -kv \ (\text{mod } \mathcal{N})$$

$$(\alpha - \mathcal{G})v - (\alpha - \mathcal{G})u = -kv \ (\text{mod } \mathcal{N})$$

$$(\alpha - \mathcal{G})v - f = -kv \ (\text{mod } \mathcal{N}), \ (\alpha - \mathcal{G})u = f \ (\text{mod } \mathcal{N})$$

$$(\alpha + k - \mathcal{G})v = f.$$

This proves the result.

As an application of Kac's theorem consider the standard Brownian motion (S,W,P_a), Let

$$\Phi(t) = \text{the Lebesgue measure of } (s : x_s > 0 \text{ and } 0 < s \leq t)$$

$$= \text{the time spent in the positive half line up to } t.$$

Note that $\Phi(t)$ is continuous in t.

Then we shall prove that

$$\frac{P_o(\Phi(t) \in d\tau)}{d\tau} = \frac{1}{\pi\sqrt{\tau(t-\tau)}}$$

so that

$$P_o(w : \Phi(t) < \tau) = \frac{2}{\pi} \text{ arc sin} \sqrt{\frac{\tau}{t}}, \quad 0 \leq \tau \leq t.$$

We have $\beta\Phi(t) = \int_o^t k(x_s)ds$ where

$$k(a) = \beta \quad \text{if} \quad a > 0$$

$$= 0 \quad \text{if} \quad a \leq 0.$$

Therefore $\beta \Phi(t) = \int_0^t k[x(s,w)]\,ds$, considered as a function of w is measurable in w. Let

$$\varphi(\beta,t,a) = E_a(e^{-\beta\Phi(t)}).$$

Then

$$\varphi(\beta,t,a) = E_a(e^{-\int_0^t k(x_s)ds})$$

$$= \int_{-\infty}^{\infty} e^{-\beta\tau} P_a(\Phi(t) \in d\tau)$$

$$= \int_0^{\infty} e^{-\beta\tau} P_a(\Phi(t) \in d\tau),$$

for, if $B \subset (-\infty, 0)$ then $P_a\{w : \Phi(t) \in B\} = 0$. Also, if

$$v(a) = v(\beta, \alpha, a) = \int_0^{\infty} e^{-\alpha t} \varphi(\beta, t, a)dt$$

we have

$$v(a) = E_a\left(\int_0^{\infty} e^{-\alpha t} f(x_t) e^{-\int_0^t k(x_s)ds} dt\right)$$

where $f \equiv 1$.

From Kac's theorem, v is a solution of the differential equation

$$(\alpha + k - \tfrac{1}{2}\frac{d^2}{da^2})y = 1 \qquad \text{(a.e.)}$$

i.e. v satisfies

$$(\alpha + \beta - \tfrac{1}{2}\frac{d^2}{da^2})y = 1 \qquad \text{if } a > 0$$

$$(\alpha - \tfrac{1}{2}\frac{d^2}{da^2})y = 1 \qquad \text{if } a < 0.$$

The general solution of this equation is

$$y = \frac{1}{\alpha+\beta} + A_1 e^{-\sqrt{2(\alpha+\beta)}x} + A_2 e^{\sqrt{2(\alpha+\beta)}x}, \qquad x > 0$$

$$= \frac{1}{\alpha} + B_1 e^{-\sqrt{2\alpha}\,x} + B_2 e^{\sqrt{2\alpha}\,x}, \qquad x < 0.$$

Since v is bounded, $A_2 = B_1 = 0$ and using the fact v is continuous and v' is continuous at 0, we have

$$v(0) = v(\beta, \alpha, 0) = \frac{1}{\sqrt{\alpha}\sqrt{\alpha+\beta}}.$$

Now,
$$\int_0^\infty e^{-\alpha t}\int_0^t \frac{1}{\sqrt{\pi(t-s)}}\frac{1}{\sqrt{\pi s}}e^{-\beta s}\,ds\,dt$$

$$= \int_0^\infty e^{-\beta s}\frac{ds}{\sqrt{\pi s}}\int_s^\infty \frac{1}{\sqrt{\pi(t-s)}}e^{-\alpha t}\,dt$$

$$= \int_0^\infty e^{-\beta s}\frac{ds}{\sqrt{\pi s}}\int_0^\infty \frac{1}{\sqrt{\pi t}}e^{-\alpha(t+s)}\,dt$$

$$= \int_0^\infty e^{-(\alpha+\beta)s}\frac{ds}{\sqrt{\pi s}}\int_0^\infty \frac{1}{\sqrt{\pi t}}e^{-\alpha t}\,dt$$

$$= \frac{1}{\sqrt{\alpha(\alpha+\beta)}}.$$

Therefore,

$$\int_0^\infty e^{-\alpha t} \varphi(\beta,t,0)dt = v(0) = \int_0^\infty e^{-\alpha t} \int_0^t \frac{1}{\sqrt{\pi(t-s)}} \frac{1}{\sqrt{\pi s}} e^{-\beta s} \, ds\, dt.$$

Fixing β , since this is true for all α , and $\varphi(\beta,t,a)$ and $\int_0^t \frac{1}{\sqrt{\pi(t-s)}} \frac{1}{\sqrt{\pi s}} e^{-\beta s} ds$ are continuous in t , we have

$$\varphi(\beta,t,0) \equiv \int_0^t \frac{1}{\sqrt{\pi(t-s)}} \frac{1}{\sqrt{\pi s}} e^{-\beta s} \, ds$$

i.e.

$$\int_0^\infty e^{-\beta\tau} P_0(\Phi(t) \in d\tau) = \int_0^t e^{-\beta s} \frac{1}{\pi\sqrt{s(t-s)}} \, ds.$$

Thus finally

$$P_0(\Phi(t) \in ds) = \frac{1}{\pi\sqrt{s(t-s)}} \, ds.$$

Section 2

Strong Markov processes

§1. Markov time

Definition. Let (S,W,P_a) be a Markov process with $W = W_{rc}$, W_{d_1} or W_c. A mapping $\sigma : W \longrightarrow [0,\infty]$ is called Markov time if

$$(w : \sigma(w) \geqslant t) \in \mathcal{B}_t .$$

It is easily seen that $w \longrightarrow w_\sigma^-$ is a measurable map of $W \longrightarrow W$. In fact, it is enough to show that

$$w \longrightarrow w_\sigma^-(t) = x(\sigma \wedge t, w)$$

is measurable, and this is immediate since $x(s,w)$, $\sigma(w)$, t and w are all measurable in the pair (s,w). Similarly, $w \longrightarrow w_\sigma^+$ is measurable.

The system of all subsets of W of the form $(w : w_{\sigma(w)}^- \in B)$, $B \in \mathcal{B}$, is denoted by $\mathcal{B}_\sigma$. $\mathcal{B}_\sigma$ is a Borel algebra contained in $\mathcal{B}$. We shall give examples to show that σ is not always $\mathcal{B}_\sigma$-measurable. However, if $\sigma < \infty$, $x_\sigma = w(\sigma(w))$ is $\mathcal{B}_\sigma$-measurable, for $x_\sigma = \lim_{t \to \infty} w_\sigma^-(t)$ and $x_\sigma(w_t^-)$ is $\mathcal{B}_\sigma$-measurable for every t.

If σ is a Markov time, then $\sigma + \epsilon$ is also a Markov time for every $\epsilon \geqslant 0$. It is not difficult to see that $\mathcal{B}_{\sigma+\epsilon}$ increases with ϵ. Let

$$\mathcal{B}_{\sigma+} = \bigcap_{\epsilon > 0} \mathcal{B}_{\sigma+\epsilon} = \bigcap_n \mathcal{B}_{\sigma+1/n} .$$

Then $\mathcal{B}_{\sigma+} \supset \mathcal{B}_\sigma$ and $\mathcal{B}_{\sigma+} \subset \mathcal{B}_{\sigma+\epsilon}$ for every $\epsilon > 0$. The class of all bounded $\mathcal{B}_\sigma$-measurable functions is denoted by $\mathfrak{B}_\sigma$ and the class of all bounded $\mathcal{B}_{\sigma+}$ -measurable functions by $\mathfrak{B}_{\sigma+}$.

Theorem. $\sigma(w)$ is $\mathcal{B}_{\sigma+}$ -measurable.

Proof, We shall prove that for every $\epsilon > 0$, $\sigma(w) = \sigma(w^-_{\sigma+\epsilon})$; from this the theorem follows. Let $w_0 \in W$. If $\sigma(w_0) = \infty$, the equality is trivial. Let $t = \sigma(w_0) < \infty$. Now

$$(w : \sigma(w) \geqslant t) \in \mathcal{B}_t \subset \mathcal{B}_{t+\epsilon}$$

for any $\epsilon > 0$. Also

$$(w : \sigma(w) > t) = \bigcup_n (w : \sigma(w) \geqslant t + \frac{\epsilon}{n}) \in \mathcal{B}_{t+\epsilon}.$$

It follows that

$$(w : \sigma(w) = t) \in \mathcal{B}_{t+\epsilon}.$$

Hence

$$(w : \sigma(w) = t) = (w : w^-_{t+\epsilon} \in B)$$

for some $B \in \mathcal{B}$. Since $\sigma(w_0) = t$ we see that $(w_0)^-_{t+\epsilon} \in B$. Hence $\left[(w_0)^-_{t+\epsilon}\right]^-_{t+\epsilon} = (w_0)^-_{t+\epsilon} \in B$. So $\sigma\left[(w_0)^-_{t+\epsilon}\right] = t$, i.e.

$$\sigma\left[(w_0)^-_{\sigma(w_0)+\epsilon}\right] = \sigma(w_0) ,$$

completing the proof.

§2. Examples of Markov time.

1. $\sigma \equiv t$.

2. $\sigma = \sigma_G = \inf. \{ t : x_t(w) \in G \}$

= first passage time for the open set $G \subset S$.

We have

$$\{ w : \sigma_G < t \} = \{ w : \exists\, s < t \text{ and } x_s \in G \}$$

$$= \{ w : \exists\, r < t \text{ , } x_r \in G \text{ and } r \text{ rational} \}$$

$$= \bigcup_{\substack{r \\ r \text{ rational, } r < t}} \{ w : x_r(w_t^-) \in G \}.$$

Thus σ_G is a Markov time.

Remark. $\sigma = \sigma_G$ is not always $\mathcal{B}_\sigma$-measurable. If σ is a Markov time which is $\mathcal{B}_\sigma$-measurable, then

$$\{ w : \sigma(w) < c \} = \{ w : w_\sigma^- \in B \} \text{ for some } B \in \mathcal{B}$$

and since $(w_\sigma^-)_\sigma^- = w_\sigma^-$, we should have $\sigma(w_\sigma^-) < c$. In particular, if σ is $\mathcal{B}_\sigma$-measurable and $\sigma(w) < \infty$, then $\sigma(w_\sigma^-) < \infty$. Now consider a Markov process with $S = (-\infty, \infty)$, $W = W_c$ and let $\sigma = \sigma_G$ where $G = (0, \infty)$. Let $w(t) = -1 + t$. Then $\sigma(w) = 1$. Also $w_\sigma^-(t) = -1+t$ if $t \leq 1$ and $w_\sigma^-(t) = 0$ if $t \geq 1$. Therefore $\sigma(w_\sigma^-) = \infty$. Hence σ cannot be $\mathcal{B}_\sigma$-measurable.

3. If $G = \{ \infty \}$, $\sigma_G = \sigma_\infty$ = killing time.

4. Let $W = W_c$ and

$$\sigma = \sigma_F = \inf.\{ t : x_t \in F \} .$$

where F is closed in S. Let $G_m \supset G_{m+1}$ be a sequence of open sets such that $\bigcap_m G_m = F$, and let $\tilde{\sigma} = \lim_m \sigma_{G_m}$. Then $\tilde{\sigma}$ is measurable [actually it is a Markov time]. We easily verify that

$$\sigma_F = \begin{cases} \tilde{\sigma} & \text{if } \tilde{\sigma} < \sigma_\infty \\ \infty & \text{if } \tilde{\sigma} = \sigma_\infty . \end{cases}$$

It follows that σ_F is measurable. Now it is easily verified that

$$[w : \sigma_F(w) < t] = [w : \sigma_F(w_t^-) < t] \quad ;$$

in fact the closedness of F is not necessary to prove this. Since $w \longrightarrow w_t^-$ is $\mathcal{B}_t$-measurable, it follows that σ_F is a Markov time.

§3. Definition of strong Markov process.

Let $\mathcal{M}$ be a Markov process. $\mathcal{M}$ is said to have the <u>strong Markov property</u> with respect to the Markov time σ if

$$P_a(w : w \in B_1, w_\sigma^+ \in B_2) = E_a(w \in B_1 : P_{x_\sigma}(B_2)) ,$$

where $B_1 \in \mathcal{B}_{\sigma+}$ and $B_2 \in \mathcal{B}$.

<u>Remark</u>. The above condition is equivalent to

$$E_a(f(w)g(w_\sigma^+)) = E_a (f(w) E_{x_\sigma}(g(w'))) ,$$

or, more generally, to

$$E_a(w \in B_1 \, , \, w^+_\sigma \in B_2 : f(w)g(w^+_\sigma)) =$$

$$= E_a(w \in B_1 : f(w) \, E_{x_\sigma}(w' \in B_2 : g(w'))) \, ,$$

where

$$f \in \mathcal{B}_{\sigma+} \, , \, g \in \mathcal{B} \, , \, B_1 \in \mathbb{B}_{\sigma+} \text{ and } B_2 \in \mathbb{B} \, .$$

Definition. $\mathbb{M}$ is called a strong Markov process if it has the strong Markov property with respect to all Markov times. A strong Markov process is called a diffusion process if $W = W_c(S)$.

§ 4. A condition for a Markov process to be a strong Markov process.

We shall later give examples to show that not all Markov processes are strong Markov processes. The following theorem gives a sufficient condition for a Markov process to be a strong Markov process.

Theorem. Let $\mathbb{M} = (S,W,P_a)$ be a Markov process and $C(S)$ the set of all real continuous bounded functions on S. If H_t maps $C(S)$ into $C(S)$, then $\mathbb{M}$ is a strong Markov process.

Proof. Let σ be a Markov time. We have to show that

$$E_a(f(w)g(w^+_\sigma)) = E_a(f(w) \, E_{x_\sigma}(g(w'))) \, .$$

Let $\delta > 0$ and $f \in \mathcal{B}_{\sigma+}$. Then, since $f \in \mathcal{B}_{\sigma+\epsilon}$ for every $\epsilon > 0$,

$$(w : t - \delta \leq f(w) < t) = (w : w^-_{\sigma+\epsilon} \in B), \; B \in \mathbb{B}.$$

Also $(w^-_{\sigma+\epsilon})^-_{\sigma+\epsilon} = w^-_{\sigma+\epsilon}$. Therefore

$$t - \delta \leq f(w^-_{\sigma+\epsilon}) < t .$$

Putting $t = f(w) + \delta$ and letting $\delta \longrightarrow 0$ we get $f(w) = f(w^-_{\sigma+\epsilon})$.

If $\sigma_m = \frac{[m\sigma]+1}{m}$, then $\sigma_m > \sigma$ and $\sigma_m \longrightarrow \sigma$ as $m \longrightarrow \infty$. We have for $f \in \mathcal{B}_{\sigma+}$ and $g_1, g_2 \in \mathcal{B}(S)$,

$$E_a(f(w)g_1(x_{t_1+\sigma})g_2(x_{t_2+\sigma})) = E_a(\sigma<\infty: f(w)g_1(x_{t_1+\sigma})g_2(x_{t_2+\sigma}))$$

and

$$f(w^-_{\sigma_m}) = f(w^-_{\sigma+\sigma_m-\sigma}) = f(w).$$

If $g_1, g_2 \in C(S)$, $g_i(x_{t_i+\sigma_m}) \longrightarrow g_i(x_{t_i+\sigma})$, i=1,2, as $m \longrightarrow \infty$.

We have therefore

$$E_a(f(w)g_1(x_{t_1+\sigma})g_2(x_{t_2+\sigma})) =$$

$$= \lim_{m\to\infty} E_a(\sigma<\infty: f(w^-_{\sigma_m})g_1(x_{t_1+\sigma_m})g_2(x_{t_2+\sigma_m}))$$

$$= \lim_{m\to\infty} \sum_{k=1}^{\infty} E_a[* : f(w^-_{\sigma_m})g_1(x_{t_1+\sigma_m})g_2(x_{t_2+\sigma_m}) ,$$

where $* \equiv (\sigma \geqslant \frac{k-1}{m}) - (\sigma \geqslant \frac{k}{m})$,

$$= \lim_{m\to\infty} \sum_{k=1}^{\infty} E_a[* : f(w^-_{k/m})g_1(x_{t_1+k/m})g_2(x_{t_2+k/m}) ,$$

since $\sigma_m = k/m$ if $\frac{k-1}{m} \leq \sigma < \frac{k}{m}$. From the definition of Markov time,

$$(\sigma \geqslant \tfrac{k-1}{m}) \in \mathcal{B}_{\frac{k-1}{m}} \subset \mathcal{B}_{\frac{k}{m}} , \quad (\sigma \geqslant k/m) \in \mathcal{B}_{k/m} ,$$

so that $* \in \mathcal{B}_{k/m}$. Therefore from the Markov property we have the last expression equal to

$$\lim_{m\to\infty} \sum_{k=1}^{\infty} E_a\Big[* : f(w^-_{k/m})E_{x_{k/m}}\big\{ g_1(x_{t_1})g_2(x_{t_2})\big\}\Big]$$

$$= \lim_{m\to\infty} \sum_{k=1}^{\infty} E_a\Big[* : f(w^-_{\sigma_m})E_{x_{\sigma_m}}\big\{g_1(x_{t_1})g_2(x_{t_2})\big\}\Big]$$

$$= \lim_{m\to\infty} E_a \Big[\sigma < \infty : f(w^-_{\sigma_m})\, F(x_{\sigma_m})\Big] ,$$

where $F(x_{\sigma_m}) = E_{x_{\sigma_m}}\big\{g_1(x_{t_1})g_2(x_{t_2})\big\}$. Also,

$$F(b) = E_b(g_1(x_{t_1})g_2(x_{t_2}))$$

$$= E_b\Big[g_1(x_{t_2}(w^-_{t_1}))g_2(x_{t_2-t_1}(w^+_{t_1}))\Big] , \text{ if } t_2 > t_1 ,$$

$$= E_b\Big[g_1(x_{t_2}(w^-_{t_1}))\ E_{x_{t_1}}(g_2(x_{t_2-t_1}(w')))\Big]$$

$$= E_b\Big[g_1(x_{t_1})H_{t_2-t_1}\ g_2(x_{t_1})\Big]$$

$$= H_{t_1}\Big[g_1\ H_{t_2-t_1}\ g_2\Big](b) .$$

Thus $F(b)$ is continuous in b since $H_t : C(S) \longrightarrow C(S)$.

Therefore

$$E_a\left[f(w)g_1(x_{t_1+\sigma})\,g_2(x_{t_2+\sigma})\right] =$$

$$= E_a\left[\sigma<\infty:\ f(w)E_{x_\sigma}(g_1(x_{t_1})g_2(x_{t_2}))\right]$$

$$= E_a\left[f(w)\ E_{x_\sigma}(g_1(x_{t_1})g_2(x_{t_2}))\right].$$

Generalizing this to $n > 2$, we have, if $g_i \in C(S)$,

$$E_a\left[f(w)g_1(x_{t_1+\sigma})\dots g_n(x_{t_n+\sigma})\right] = E_a\left[f(w)E_{x_\sigma}(g_1(x_{t_1})\dots g_n(x_{t_n}))\right].$$

The same equation holds if $g_i \in \mathcal{B}(S)$. If $B \in \mathbb{B}$ and $B = (w : w(t_1) \in E_1,\dots,w(t_n) \in E_n)$, then

$$\chi_B(w) = \chi_{E_1}(x_{t_1}) \dots \chi_{E_n}(x_{t_n}),$$

and therefore,

$$E_a[f(w)\chi_B(w^+_\sigma)] = E_a\left[f(w)E_{x_\sigma}(\chi_{E_1}(x_{t_1}) \dots \chi_{E_n}(x_{t_n})\right.$$

$$= E_a\left[f(w)E_{x_\sigma}(\chi_B(w'))\right].$$

The equation

$$E_a(f(w)g(w^+_\sigma)) = E_a\left[f(w)E_{x_\sigma}(g(w'))\right]$$

follows easily now for $g \in \mathcal{B}$.

§5. <u>Example of a Markov process which is not a strong Markov process.</u>

The above theorem shows that all the preceding examples of

Markov processes are strong Markov processes. The natural question is whether there exist Markov processes which are not strong Markov processes. The following example answers this question in the affirmative.

Suppose that $\Omega(P)$ is a probability space and $\tau(w)$, $w \in \Omega$, a random variable on $\Omega(P)$ such that

$$P(w : \tau(w) \in E) = \int_E \lambda e^{-\lambda t}\, dt\ , \quad \lambda > 0 .$$

Such a random variable is often called <u>exponential holding time</u>.

Let $S = [0, \infty)$ and $W = W_c$. Define

$$\xi^{(a)}(t,w) = a + t\ , \quad a > 0 ;$$

$$\xi^{(0)}(t,w) = \begin{cases} 0\ , & \text{if } t < \tau(w)\ , \\ t - \tau\ , & \text{if } t \geq \tau(w). \end{cases}$$

$\xi^{(a)}(t,w)$ are random variables on $\Omega(P)$ and for fixed w are in W_c. For $B \in \mathcal{B}(W)$ and $0 \leq a < \infty$, define

$$P_a(B) = P\left[w : \xi^{(a)}(.,w) \in B\right].$$

For $a > 0$, then, $P_a(B) = 1$ if $\xi^{(a)} \in B$ and $P_a(B) = 0$ otherwise.

To show that $\mathcal{M} = (S,W,P_a)$ is a Markov process, we have only to verify the Markov property. To do this, we show that if f_1, $f_2 \in \mathcal{B}(S)$, then

$$E_a(f_1(x_{t_1})f_2(x_{t_2})) = H_{t_1}(f_1 H_{t_2-t_1} f_2)(a).$$

Denoting by E the expectation on Ω, we have

$$H_t f(a) = f(a+t),\ a > 0 ;$$

$$H_t f(0) = E_o(f(x_t)) = E(\tau \leq t; f(t-\tau)) + E(t < \tau;\ f(0)).$$

So if $a > 0$,

$$\begin{aligned} E_a(f_1(x_{t_1})f_2(x_{t_2})) &= f_1(a+t_1)f_2(a+t_2). \\ &= f_1(a+t_1)H_{t_2-t_1} f(a+t_1) \\ &= H_{t_1}(f_1 H_{t_2-t_1} f_2)(a) \end{aligned}$$

If $a = 0$, we have

$$\begin{aligned} E_o(f_1(x_{t_1})f_2(x_{t_2})) &= E(\tau \leq t_1; f_1(t_1-\tau)f_2(t_2-\tau)) + \\ &\quad + E(t_1 < \tau \leq t_2; f_1(0)f_2(t_2-\tau)) + E(t_2 < \tau:\ f_1(0)f_2(0)) \\ &= \int_0^{t_1} f_1(t_1-s)f_2(t_2-s)\lambda e^{-\lambda s} ds + f_1(0)\int_{t_1}^{t_2} f_2(t_2-s)\lambda e^{-\lambda s} ds + f_1(0)f_2(0)e^{-\lambda t_2} \\ &= \int_0^{t_1} f_1(t_1-s)f_2(t_2-s)\lambda e^{-\lambda s} ds + f_1(0)e^{-\lambda t_1}\int_0^{t_2-t_1} f_2(t_2-t_1-s)\lambda e^{-\lambda s} ds + \\ &\quad + f_1(0)f_2(0)\ e^{-\lambda t_2} \\ &= \int_0^{t_1} f_1(t_1-s)\left[H_{t_2-t_1} f_2(t_1-s)\right] \lambda e^{-\lambda s} ds + \\ &\quad + f_1(0)e^{-\lambda t_1}\left[\int_0^{t_2-t_1} f_2(t_2-t_1-s)\lambda e^{-\lambda s} ds + f_2(0)e^{-\lambda(t_2-t_1)}\right] \end{aligned}$$

$$= \int_0^{t_1} f_1(t_1-s)H_{t_2-t_1}f_2(t_1-s)\lambda e^{-\lambda s} + f_1(0)e^{-\lambda t_1}H_{t_2-t_1}f_2(0)$$

$$= H_{t_1}\left[f_1 \, H_{t_2-t_1} \, f_2\right](0).$$

The following facts are easily verified:

$$\mathcal{N} = \left\{ f : f = 0 \text{ a.e., } f(0) = 0 \right\} ;$$

$$\mathcal{R} = \left\{ u : u \text{ abs. cont. in } (0, \infty), \ u, u' \in \mathcal{B}(0, \infty) \right\};$$

$\mathcal{G} u(a) = u'(a)$ for $a > 0$ and $\mathcal{G} u(0) = [u(0+)-u(0)]\lambda$.

We now show that $\mathcal{M}$ is <u>not</u> a strong Markov process. Let $\sigma = \sigma_G$, where $G = (0, \infty)$. We shall show that $\mathcal{M}$ does not have the strong Markov property with respect to the Markov time σ. We have,

$$A = P_0(\sigma > 0, \ \sigma(W_\sigma^+) > 0) = 0 ,$$

since $\sigma(w_\sigma^+) = 0$. Also

$$\left\{ w : \tau(w) > 0 \right\} \subset \left\{ w : \sigma(\xi^{(0)}(.,w)) > 0 \right\},$$

and hence

$$P_0(w : \sigma(w) > 0) = P(w : \sigma(\xi^{(0)}(.,w)) > 0) \geq P\left\{ w : \tau(w) > 0 \right\} = 1.$$

Note that $w(\sigma(w)) = 0$. If $\mathcal{M}$ has the strong Markov property with respect to σ, we should have

$$0=A=P_0(\sigma>0,\sigma(w_\sigma^+)>0)=E_0(\sigma>0;P_{x_\sigma}(\sigma>0))=E_0(\sigma>0;P_0(\sigma>0))=1\cdot P_0(\sigma>0)=1,$$

but this is absurd.

§6. Dynkin's formula and generalized first passage time relation.

We now prove some theorems on Markov processes which have the strong Markov property with respect to the Markov time σ.

Theorem 1 (Dynkin). If $u(a) = G_\alpha f(a)$, then

$$u(a) = E_a\left(\int_0^\sigma e^{-\alpha t} f(x_t)dt\right) + E_a(e^{-\alpha\sigma}u(x_\sigma)).$$

Proof. We have

$$\begin{aligned} u(a) &= E_a\left(\int_0^\infty e^{-\alpha t} f(x_t)dt\right) \\ &= E_a\left(\int_0^\sigma e^{-\alpha t} f(x_t)dt\right) + E_a\left(\int_\sigma^\infty e^{-\alpha t} f(x_t)dt\right), \end{aligned}$$

and

$$\begin{aligned} E_a\left(\int_\sigma^\infty e^{-\alpha t} f(x_t)dt\right) &= E_a\left(e^{-\alpha\sigma}\int_0^\infty e^{-\alpha t} f(x_t(w_\sigma^+))dt\right) \\ &= \int_0^\infty e^{-\alpha t} E_a(e^{-\alpha\sigma} f(x_t(w_\sigma^+)))dt \\ &= \int_0^\infty e^{-\alpha t} E_a(e^{-\alpha\sigma} E_{x_\sigma}(f(x_t))dt, \end{aligned}$$

because $\mathcal{M}$ has the strong Markov property with respect to σ. (Note

that if φ is a Borel function on the real line, then $\varphi(\sigma) \in \mathcal{B}_{\sigma+}$).
Therefore

$$E_a\left(\int_\sigma^\infty e^{-\alpha t} f(x_t)dt\right) = E_a\left(\int_0^\infty e^{-\alpha\sigma} E_{x_\sigma}(e^{-\alpha t} f(x_t))dt\right)$$

$$= E_a(e^{-\alpha\sigma} u(x_\sigma)).$$

Before proving Theorem 2, we prove the following

Lemma. Let

$$\mu_a(dt\ db) = P_a[\sigma \in dt,\ x_\sigma \in db]$$

be the measure induced on the Borel sets of $R^1 \times S$ by the mapping $w \longrightarrow (\sigma, x_\sigma)$ of W into $R^1 \times S$. Let $\varphi(t,b)$ be a bounded Borel measurable function on $R^1 \times S$. Then

$$\int_{[0,\infty)\times S} e^{-\alpha t} \mu_a(dt\ db) \int_{s=0}^\infty e^{-\alpha s} \varphi(s,b)ds$$

$$= \int_0^\infty e^{-\alpha t}\ dt \int_{[0,t]\times S} \varphi(t-s,b)\mu_a(ds\ db)$$

Proof. We have

$$\int_{[0,\infty)\times S} e^{-\alpha t} \mu_a(dt\ db) \int_0^\infty e^{-\alpha s} \varphi(s,b)ds = \int_{[0,\infty)\times S} \mu_a(dt\ db) \int_t^\infty e^{-\alpha s} \varphi(s-t,b)ds$$

$$= \int_{[0,\infty)\times S} \mu_a(dt\ db) \int_0^\infty F(t,s,b)ds$$

where

$$F(t,s,b) = e^{-\alpha s}\varphi(s-t,b) \ , \text{ if } \ s \geqslant t \ ;$$

$$0 \ , \text{ if } \ s < t.$$

Changing the order of integration, we get the last expression equal to

$$\int_0^\infty ds \int_{[0,\infty)\times S} F(t,s,b)\,\mu_a(dt\ db)$$

$$= \int_0^\infty e^{-\alpha s}\, ds \int_{[0,s]\times S} \varphi(s-t,b)\,\mu_a(dt\ db).$$

This proves the lemma.

<u>Theorem 2</u> (Generalized first passage time relation). Put

$$Q(t,a,E) = P_a(x_t \in E \ \text{ and } \ \sigma > t \ ).$$

Then

$$P(t,a,E) = Q(t,a,E) + \int_{[0,t]\times S} P(t-s,b,E)\,\mu_a\ (ds\ db)$$

<u>Remark</u>. When σ is the first passage time, this is usually known as the 'first passage time relation'.

<u>Proof</u>. We have

$$E_a\left(\int_0^\sigma e^{-\alpha t} f(x_t)dt\right) = E_a\left(\int_0^\infty e^{-\alpha t} f(x_t)\,\chi_{[0,\sigma]}(t)dt\right)$$

$$= \int_0^\infty e^{-\alpha t}\, E_a(\sigma > t : f(x_t))dt \ .$$

Further

$$E_a(e^{-\alpha\sigma}u(x_\sigma)) = \int_{[0,\infty)\times S} e^{-\alpha t}\, u(b)\,\mu_a(dt\, db) ,$$

and since $u(b) = \int_0^\infty e^{-\alpha s} H_s f(b)ds$, we have from the Lemma,

$$E_a(e^{-\alpha\sigma}u(x_\sigma)) = \int_{t=0}^\infty e^{-\alpha t}\, dt \int_{[0,t]\times S} H_{t-s}f(b)\,\mu_a(ds\, db).$$

From Theorem 1, therefore

$$\int_0^\infty e^{-\alpha t} H_t f(a)dt = u(a) = \int_0^\infty e^{-\alpha t} E_a(\sigma > t : f(x_t))dt +$$

$$+ \int_{t=0}^\infty e^{-\alpha t}\, dt \int_{[0,t]\times S} H_{t-s}f(b)\,\mu_a(ds\, db).$$

Since the last equation is true for all $\alpha > 0$, we have for almost all t ,

$$H_t f(a) = E_a(\sigma > t;\, f(x_t)) + \int_{[0,t]\times S} H_{t-s}f(b)\,\mu_a(dsdb).$$

Now suppose that f is bounded and continuous. Then

$$H_t f(a) - E_a(\sigma > t : f(x_t)) = E_a(\sigma \le t; f(x_t)) = E_a(f(x_t)\,\chi_{[0,t]}(\sigma(w)))$$

is right continuous in t since $f(x_t)$ and $\chi_{[0,t]}$ are right continuous in t . Further

$$\int_{[0,t]\times S} H_{t-s}f(b)\,\mu_a(ds\,db) = \int_{[0,\infty]\times S} \chi_{[0,t]}(s)H_{t-s}f(b)\,\mu_a(ds\,db)$$

and so is also right continuous in t . Therefore the above equation holds for all t if f is continuous and bounded. It follows easily that for any $f \in \mathcal{B}(S)$ the equation is true identically in t . Putting $f = \chi_E$, we get Theorem 2.

The following rough proof should give us an intuitive explanation of Theorem 2.

$$P(t,a,E) - Q(t,a,E) = P_a(x_t \in E,\ t \geq \sigma)$$

$$= \int_{s=0}^{t} \int_S P_a(\sigma \in ds,\ x_\sigma \in db\ ,\ x_t \in E)$$

$$= \int_{s=0}^{t} \int_S P_a(\sigma \in ds,\ x_s \in db\ ,\ x_{t-s}(w_s^+) \in E)$$

$$= \int_{s=0}^{t} \int_S P_a(\sigma \in ds,\ x_s \in db)\ P_b(x_{t-s} \in E)$$

$$= \int_{s=0}^{t} \int_S P(t-s,b,E)\ \mu(ds\ db).$$

We give below two examples to illustrate the use of Theorem 2.

Example 1. Let M be the standard Brownian motion, $E \in \mathcal{B}(0,\infty)$ and $a > 0$. Then we shall prove that

$$P_a(x_t \in E,\ t < \sigma_o) = \int_E [N(t,a,b) - N(t,a,-b)]\, db$$

where $\sigma_o(w) = \inf\ (t : w(t) = 0)$.

Since $w(\sigma_o(w)) = 0$, for $E \in \mathcal{B}[0,\infty)$ and $F \in \mathcal{B}(S)$ we have

$$\mu_a(E \times F) = P_a(\sigma_o \in E,\ x_{\sigma_o} \in F) = 0\ , \text{ if } 0 \notin F\ ;$$

$$P_a(\sigma_o \in E), \text{ if } 0 \in F.$$

Therefore from Theorem 2, with $\sigma = \sigma_o$, we have

$$P(t,a,E) = Q(t,a,E) + \int_{s=o}^{t} P(t-s,o,E)\ \mu_a(ds)\ ,$$

$$P(t,a,-E) = Q(t,a,-E) + \int_{s=o}^{t} P(t-s,o,-E)\ \mu_a(ds).$$

Since $a > o$, $E \in \mathcal{B}(o,\infty)$ and all continuous paths starting at a and going into $-E$ pass through o , $Q(t,a,-E) = 0$. Also $P(t-s,o,E) = P(t-s,o,-E)$. Therefore, subtracting,

$$P(t,a,E)-P(t,a,-E) = Q(t,a,E) = P_a(x_t \in E\ ,\ t < \sigma_o)$$

i.e. $$\int_E [N(t,a,b) - N(t,a,-b)]\, db = P_a(x_t \in E,\ t < \sigma_o)\ .$$

<u>Remark</u>. $P_a(x_t \in E$ and $t < \sigma_o) = P_a(x_t \in E$ and $x_s > o\ ,\ o \leq s \leq t\ )$.

<u>Example 2</u>.

$$P_a(x_s > o,\ o \leq s \leq t) = 2 \int_o^a \frac{1}{\sqrt{2\pi t}}\, e^{-\xi^2/2t}\, d\xi = P_o(|x_t| < a\ ).$$

Put $E = (0,\infty)$ in Example 1. Then we get

$$P_a(x_s > 0,\ 0 \leq s \leq t) = \int_0^{\infty} \frac{1}{\sqrt{2\pi t}} \left(e^{-\frac{(b-a)^2}{2t}} - e^{-\frac{(b+a)^2}{2t}}\right) db$$

$$= 2 \int_0^{a} \frac{1}{\sqrt{2\pi t}}\, e^{-\xi^2/2t}\, d\xi$$

$$= P_0(|x_t| < a).$$

Note that if $a > 0$,

$$P_a(x_s > 0, 0 \leq s \leq t) = P_a(\sigma_0 > t) =$$

$$= P_a(\min_{0 \leq s \leq t} x_s > 0) = P_a(\min_{0 \leq s \leq t} x_s > -a)$$

$$= P_0(\max_{0 \leq s \leq t} x_s < a).$$

The following important theorem which follows easily from Theorem 1 gives what is called Dynkin's formula.

<u>Theorem 3</u>. If $E_a(\sigma) < \infty$ and $u \in \mathcal{D}(\mathcal{G})$, then

$$E_a\left(\int_0^{\sigma} \mathcal{G}u(x_t)dt\right) = E_a(u(x_\sigma)) - u(a).$$

<u>Proof</u>. From Theorem 1, **If** $u = G_\alpha f$, **then**

$$u(a) = E_a\left(\int_0^{\sigma} e^{-\alpha t} f(x_t)dt\right) + E_a(e^{-\alpha\sigma} u(x_\sigma)).$$

Also $$f(x_t) = \alpha u(x_t) - \mathcal{G}u(x_t).$$

Therefore

$$u(a) = E_a\left\{\int_0^{\sigma} e^{-\alpha t}(\alpha u(x_t) - \mathcal{G}u(x_t))dt\right\} + E_a(e^{-\alpha\sigma}u(x_{\sigma})).$$

Letting $\alpha \longrightarrow 0$, we get the result.

§ 7. Blumenthal's 0-1 law.

Let $\mathcal{M}$ denote a strong Markov process.

Theorem 1. If $A \in \mathcal{B}_{0+}$ $(= \bigcap_{\epsilon > 0} \mathcal{B}_{\epsilon})$, then $P_a(A) = 1$ or 0.

Proof. For $P_a(A) = P_a(A, w \in A) = P_a(A, w_0^+ \in A)$

$$= E_a(A : P_{x_0}(A)) = E_a(P_a(A) : A) = (P_a(A))^2.$$

Theorem 2. If $f(w) \in \mathcal{B}_{0+}$, then $P_a(f = E_a(f)) = 1$.

Proof. Since $f \in \mathcal{B}_{0+}$, f is bounded. From Theorem 1,

$$P_a[f > E_a(f)] = 1 \text{ or } 0.$$

Obviously it cannot be 1, since then $E_a(f) < E_a(f)$. Hence $P_a[f > E_a(f)] = 0$. For the same reason, $P_a[f < E_a(f)] = 0$. Hence $P_a[f = E_a(f)] = 1$.

We consider the following

Example. Let $\varphi(t)$ be a function of t , positive and increasing for $t > 0$. Let x_t be a real valued strong Markov process. Consider

$$p_a(\varphi) = P_a\left[\lim_{\delta\downarrow 0}\bigcap_{0\le t\le\delta}(|x_t - a| \le \varphi(t))\right].$$

By Theorem 1, $p_a(\varphi) = 1$ or 0. If $p_a(\varphi) = 1$, we say that $\varphi \in \mathcal{U}_a$ (the upper class) and if $p_a(\varphi) = 0$, we say that $\varphi \in \mathcal{L}_a$ (the lower class). Wiener proved that, for the Brownian motion,

$$\varphi(t) = t^{\frac{1}{2}-\varepsilon} \in \mathcal{U}_a \text{ and } \varphi(t) = t^{\frac{1}{2}+\varepsilon} \in \mathcal{L}_a \text{ for every } \varepsilon > 0.$$

These results have been made more precise by P. Levy, Kolmogoroff and Erdös. P. Lévy's theorem is that

$$\varphi(t) = (1+c)\sqrt{2t\log\log 1/t} \in \begin{cases}\mathcal{U}_a, & c>0\\ \mathcal{L}_a, & c<0.\end{cases}$$

§ 8. Markov process with discrete state space.

Let $\mathbb{M}$ be a right continuous Markov process with discrete state space S. Since S satisfies the second countability axiom, it is countable. We denote the elements of S by (1,2,3,...). Since S is discrete, $\mathcal{B}(S) = C(S)$ and W consists of the set of all step-functions before their killing-time. $\mathbb{M}$ is a strong Markov process because $H_t C(S) \subset C(S)$.

Let $\tau_a = \inf(t : x_t \neq a) = \inf(t : x_t \in G)$ where $G = (S - \{a\}) \cup \{\infty\}$. τ_a is called the first leaving time from a. Clearly $\tau_a \le \sigma_\infty$. τ_a has the following properties:

1. τ_a is a Markov time.

For, $(\tau_a \geq t) = (x_s = a \text{ for all } s < t)$

$$= (x_r = a \text{ for all } r < t,\ r \text{ rational}) \in \mathcal{B}_t .$$

Note that

$$(\tau_a > t) = (x_s = a \text{ for all } s \leq t)$$

$$= (x_r = a \text{ for all rational } r < t \text{ and } x_t = a) \in \mathcal{B}_t .$$

2. $P_a(\tau_a > t) = e^{-p_a t}$ where $\frac{1}{p_a} = E_a(\tau_a)$.

Indeed we have

$$P_a(\tau_a > t+s) = P_a(\tau_a > t, \tau_a(w_t^+) > s)$$

$$= E_a(\tau_a > t, P_{x_t}(\tau_a > s))$$

$$= E_a(\tau_a > t, P_a(\tau_a > s)), \text{ since } x_t = a ,$$

$$= P_a(\tau_a > t)\, P_a(\tau_a > s)$$

Therefore, if $\varphi(t) = P_a(\tau_a > t)$, then $\varphi(t)$ is right continuous, as is easily seen; $0 \leq \varphi(t) \leq 1$ and $\varphi(t+s) = \varphi(t)\,\varphi(s)$. Further

$$\varphi(0) = P_a(\tau_a > 0) = P_a(w : x_o = a) = 1.$$

If $\varphi(t) = 0$ for some $t > 0$, then $\varphi(t) = (\varphi(t/n))^n = 0$ and so we should have $\varphi(t/n) = 0$ for all n, and by right continuity,

$\varphi(0) = 0$. Therefore $0 < \varphi(t) \leq 1$ for all t . Thus

$$\varphi(t) = e^{-p_a t}, \quad 0 \leq p_a < \infty .$$

If $p_a = 0$, then $\varphi(t) \equiv 1$, i.e. $P_a(\tau_a > t) = 1$, i.e. $P_a(\tau_a = \infty) = 1$ and so

$$E_a(\tau_a) = \int_{\tau_a = \infty} \tau_a(w)dP_a(w) = \infty .$$

If $p_a > 0$, the map $w \longrightarrow \tau_a(w)$ induces the measure $p_a e^{-p_a t} dt$.
Therefore

$$E_a(\tau_a) = \int_0^\infty t p_a e^{-p_a t} dt = \frac{1}{p_a} .$$

3. x_{τ_a} and τ_a are independent with respect to P_a.

Indeed, noticing that $\tau_a(w) = t + \tau_a(w_t^+)$ if $\tau_a(w) > t$, we have

$$\begin{aligned} P_a(\tau_a > t, x_{\tau_a} \in E) &= P_a(\tau_a > t, x_{t+\tau_a(w_t^+)}(w) \in E) \\ &= P_a(\tau_a > t, x_{\tau_a(w_t^+)}(w_t^+) \in E) \\ &= E_a(\tau_a > t, P_{x_t}(x_{\tau_a} \in E)) \\ &= E_a(\tau_a > t, P_a(x_{\tau_a} \in E)) \\ &= P_a(x_{\tau_a} \in E)\, P_a(\tau_a > t). \end{aligned}$$

We now determine the generator. From Theorem 1 of §6, if $u = G_\alpha f$,

$$u(a) = E_a\left(\int_0^{\tau_a} f(x_t)e^{-\alpha t}\,dt\right) + E_a\left(e^{-\alpha\tau_a}u(x_{\tau_a})\right)$$

Since $w(t) = a$ for $t < \tau_a(w)$ and τ_a, x_{τ_a} are independent, we have

$$u(a) = E_a\left(f(a)\int_0^{\tau_a} e^{-\alpha t}\,dt\right) + E_a\left(e^{-\alpha\tau_a}\right)E_a\left(u(x_{\tau_a})\right)$$

$$= f(a)E_a\left(\frac{1-e^{-\alpha\tau_a}}{\alpha}\right) + E_a\left(e^{-\alpha\tau_a}\right)E_a\left(u(x_{\tau_a})\right)$$

$$= f(a)\int_0^{\infty}\frac{1-e^{-\alpha t}}{\alpha}\,e^{-p_a t}p_a\,dt + E_a\left(u(x_{\tau_a})\right)\int_0^{\infty} e^{-\alpha t}e^{-p_a t}p_a\,dt$$

$$= \frac{f(a)}{p_a+\alpha} + \frac{p_a}{p_a+\alpha}E_a\left(u(x_{\tau_a})\right).$$

Let now

$$\pi_{ab} = P_a(x_{\tau_a} = b).$$

Then

$$E_a\left(u(x_{\tau_a})\right) = \sum_{b\in S\cup\infty}\pi_{ab}\,u(b).$$

Since $u(\infty)$ is by definition zero,

$$u(a) = \frac{f(a)}{p_a+\alpha} + \frac{p_a}{p_a+\alpha}\sum_{b\in S}\pi_{ab}u(b).$$

From the last equation we see that $u \equiv 0$ implies $f \equiv 0$. Therefore

$$\mathcal{N} = \{f : f \equiv 0\}.$$

Also from the above we get

$$\alpha u(a) - f(a) = p_a \sum_{b \in S} \pi_{ab} u(b) - p_a u(a)$$

and hence

$$\mathcal{G} u(a) = p_a (\sum_{b \in S} \pi_{ab} u(b) - u(a))$$

$$= p_a (\sum_{b \in S} \pi_{ab} (u(b) - u(a)) - \pi_{a\infty} u(a))$$

since $\sum_{b \in S} \pi_{ab} + \pi_{a\infty} = 1$.

__Remark.__ It is generally difficult to determine $\mathcal{R} = \mathcal{D}(\mathcal{G})$.

We can also find $\mathcal{G}$ from Dynkin's formula as follows:

$$E_a(\int_0^{\tau_a} \mathcal{G} u(x_t) dt) = E_a(u(x_{\tau_a})) - u(a).$$

Therefore

$$\mathcal{G} u(a) \quad E_a[\int_0^{\tau_a} \cdots dt] = \sum_{b \in S} \pi_{ab} u(b) - u(a)$$

i.e. $\mathcal{G} u(a) E_a(\tau_a) = \sum_{b \in S} \pi_{ab} u(b) - u(a)$

and since $E_a(\tau_a) = 1/p_a$, we get the result.

Example. Suppose that $\pi_{ab} = 0$ except for $b = a\pm1$ or $b = \infty$ and let

$$\pi_{a,a+1} = \mu_a \ , \ \pi_{a,a-1} = \nu_a \ , \ \pi_{a\infty} = \lambda_a = 1 - \mu_a - \nu_a \ .$$

This process is called the <u>birth and death process</u>. We have

$$\begin{aligned} \mathcal{G}u(a) &= p_a(\mu_a u(a+1) + \nu_a u(a-1) - u(a)) \\ &= p_a\left[\mu_a(u(a+1)-u(a)) + \nu_a(u(a-1)-u(a)) - \lambda_a u(a)\right] \end{aligned}$$

In this particular case we can determine $\mathcal{D}\mathcal{G}$ which will depend on the behaviour of p_a , μ_a and ν_a at $a = \infty$.

§9. Generator in the restricted sense.

In the case of the generator $\mathcal{G}$ defined previously there was some ambiguity so that $\mathcal{G}u(a)$ had no meaning unless we took a version of $\mathcal{G}u$. We shall now avoid this ambiguity by restricting the domain of the generator; we can then speak of $\mathcal{G}u(a)$. Before doing this we prove some theorems on the domain of the new generator. We first define the function space $\mathcal{D}(S)$.

Definition. Let y_t , $t > 0$, be a random process on a probability space $\Omega(\mathcal{B},P)$. We say that y_t <u>tends to y essentially (P)</u> as $t \downarrow t_0$, in symbols: $y_t \xrightarrow[\text{ess.}(P)]{} y$, if for any countable t - set C with $t_0 \in \bar{C}$,

$$P(\lim_{t\in C, t\to t_0} y_t = y) = 1.$$

Let $\mathcal{M} = (S,W,P_a)$ be a strong Markov process. We make the following

<u>Definition</u>.

$$\mathcal{D}(S) = \left\{f : f \in \mathcal{B}(S) \text{ and for every } a,\ f(x_t) \xrightarrow[\text{ess.}(P_a)]{} f(a), \text{as } t \downarrow 0\right\}.$$

<u>Theorem 1</u>. $\mathcal{D}(S) \supset C(S)$.

<u>Proof</u>. Clear.

<u>Theorem 2</u>. $G_\alpha \mathcal{B}(S) \subset \mathcal{D}(S)$. In particular, $G_\alpha \mathcal{D}(S) \subset \mathcal{D}(S)$.

The proof depends on the following Lemma, the proof of which can be seen in Doob's book (p.355).

<u>Lemma</u>. Let z be a random variable on a probability space $\Omega(\mathcal{B},P)$, with $E(|z|) < \infty$. Let $\mathcal{B}_t \subset \mathcal{B}$, $0 < t < \infty$, be Borel algebras such that if $t < s$, $\mathcal{B}_t \subset \mathcal{B}_s$. Then, if $\mathcal{B}_{0+} = \bigcap_{t>0} \mathcal{B}_t$, we have

$$E(z/\mathcal{B}_t) \xrightarrow[\text{ess. }(P)]{} E(z/\mathcal{B}_{0+}).$$

<u>Proof of Theorem 2</u>: We prove first that

$$G_\alpha f(x_t) = e^{\alpha t} E_a(z/\mathcal{B}_t) - e^{\alpha t} \int_0^t e^{-\alpha s} f(x_s)ds$$

with P_a probability 1, where $z = \int_0^\infty e^{-\alpha s} f(x_s)ds$. Indeed, if $B_t \in \mathcal{B}_t$, by the Markov property,

$$E_a(G_\alpha f(x_t):B_t) = E_a(E_{x_t}(\int_0^\infty e^{-\alpha s} f(x_s)ds):B_t)$$

$$= E_a(\int_0^\infty e^{-\alpha s} f(x_s(w_t^+))ds:B_t)$$

$$= e^{\alpha t} E_a(\int_t^\infty e^{-\alpha s} f(x_s)ds:B_t).$$

Since $G_\alpha f(x_t) \in \mathcal{B}_t$, by the definition of conditional expectation we have

$$G_\alpha f(x_t) = e^{\alpha t} E_a(\int_t^\infty e^{-\alpha s} f(x_s)ds / \mathcal{B}_t)$$

$$= e^{\alpha t} E_a(\int_0^\infty e^{-\alpha s} f(x_s)ds / \mathcal{B}_t) - e^{\alpha t} E_a(\int_0^t e^{-\alpha s} f(x_s)ds / \mathcal{B}_t)$$

$$= e^{\alpha t} E_a(z | \mathcal{B}_t) - e^{\alpha t} (\int_0^t e^{-\alpha s} f(x_s)ds).$$

Since $\int_0^t e^{-\alpha s} f(x_s)ds \in \mathcal{B}_t$, the conditional expectation of $\int_0^t e^{-\alpha s} f(x_s)ds$ is $\int_0^t e^{-\alpha s} f(x_s)ds$ with probability 1. Using the Lemma, therefore,

$$G_\alpha f(x_t) \xrightarrow[\text{ess } (P_a)]{} E_a(z / \mathcal{B}_{0+}) \text{ as } t \downarrow 0 .$$

From Blumenthal's 0-1 law, if $E \in \mathcal{B}_{0+}$, $P_a(E) = 0$ or 1. Hence

$$E_a(z / \mathcal{B}_{0+}) = E_a(z) = G_\alpha f(a).$$

This proves the theorem.

<u>Theorem 3</u>. If $f \in \mathcal{D}(S)$, $f(x_t)$ is right continuous with respect to L^1-norm.

<u>Proof</u>. Since $f \in \mathcal{D}(S)$, if $t_n \longrightarrow 0$, $P_a(f(x_{t_n}) \longrightarrow f(a)) = 1$, so that

$$E_a(|f(x_{t_n}) - f(a)|) \longrightarrow 0 \quad \text{as} \quad n \longrightarrow \infty .$$

Now

$$E_a(|f(x_{s+t})-f(x_s)|) = E_a(|f(x_t(w_s^+)) - f(x_o(w_s^+))|)$$

$$= E_a(E_{x_s}(|f(x_t)-f(x_o)|)) \longrightarrow 0 \text{ as } t \longrightarrow 0$$

This proves the result.

<u>Theorem 4</u>. If $f \in \mathcal{D}(S)$ and $G_\alpha f = 0$, then $f \equiv 0$.

<u>Proof</u>. Note that if $G_\alpha f = 0$ for some β, $G_\beta f = 0$ for all β, from the resolvent equation. From Theorem 3,

$$H_t f(a) = E_a(f(x_t)) \longrightarrow f(a) \text{ as } t \longrightarrow 0.$$

Now $$0 = \alpha G_\alpha f(a) = \alpha \int_o^\infty e^{-\alpha t} H_t f(a)dt$$

$$= \int_o^\infty e^{-s} H_{s/\alpha} f(a)ds \longrightarrow f(a) \text{ as } \alpha \longrightarrow \infty. \quad \text{Q.E.D.}$$

<u>Theorem 5</u>. If $f \in \mathcal{D}(S)$,

$$P_a\left(\frac{1}{t} \int_0^t f(x_s)ds \longrightarrow f(a) \text{ as } t \longrightarrow 0\right) = 1.$$

<u>Proof</u>. Put $y(s,w) = f(x_s(w)) - f(a)$ and let $C = \{ 2^{-n}k,\ k,n=1,2,\ldots\}$ be the set of dyadic rational numbers. Then from the definition of $\mathcal{D}(S)$,

$$\lim_{t \to 0} \sup_{s \in C, 0 \leq s \leq t} |y(s,w)| = 0 ,$$

for $w \in \Omega_1$, with $P_a(\Omega_1) = 1$. Put $\varphi_n(s) = \dfrac{[2^n s]+1}{2^n}$. Then $\varphi_n(s) \longrightarrow s$ for every s. From Theorem 3,

$$\int_0^1 E_a(|y(\varphi_n(s),w) - y(s,w)|)ds \longrightarrow 0 \text{ as } n \longrightarrow \infty,$$

i.e. $y(\varphi_n(s),w) \longrightarrow y(s,w)$ in L^1-norm on $L^1([0,1] \times W)$. Therefore, there exists a subsequence, $\psi_n(s) = \varphi_{k_n}(s)$, say, such that $y(\psi_n(s),w) \longrightarrow y(s,w)$ for $(s,w) \in A$, say, with $(m \times P_a)(A) = 1$, $m \times P_a$ denoting the product measure on $[0,1] \times W$. Now

$$(m \times P_a)(A) = \int m(s:(s,w) \in A)dP_a(w) = 1 ,$$

so that $m(s:(s,w) \in A) = 1$ for $w \in \Omega_2$, $P_a(\Omega_2) = 1$. Let $\Omega_1 \cap \Omega_2 = \Omega$. Then if $w \in \Omega$, $w \in \Omega_2$ so that

$$\left| \frac{1}{t} \int_0^t y(s,w)ds \right| = \lim_n \left| \frac{1}{t} \int_0^t y(\psi_n(s),w)ds \right|$$

$$\leq \lim_n \sup_{s \in C, 0 \leq s \leq t} y(s,w) \rightarrow 0 \text{ as } t \rightarrow 0 ,$$

since $w \in \Omega_1$.

<u>Definition of generator in the restricted sense</u>. Let $\mathcal{M}$ be a strong Markov process. Consider the restriction of G_α to $\mathcal{D}(S)$. We shall denote this also by G_α.

<u>Theorem 6</u>. $\mathcal{R}_\alpha = G_\alpha \mathcal{D}(S)$ is independent of α. (We can therefore denote $\mathcal{R}_\alpha$ by $\mathcal{R}$.)

The proof is similar to that is the case of the generator defined earlier.

<u>Theorem 7</u>. $G_\alpha : \mathcal{D}(S) \longrightarrow \mathcal{R}$ is 1:1 and linear.

<u>Proof</u>. Since $G_\alpha f = 0$ implies $f \equiv 0$, G_α is 1:1.

Let us write $\mathcal{G}_\alpha = \alpha - G_\alpha^{-1}$.

<u>Theorem 8</u>. $\mathcal{G}_\alpha$ is independent of α.

This is obvious.

<u>Definition</u>. $\mathcal{G} = \alpha - G_\alpha^{-1}$ is called the <u>generator in the restricted sense</u>.

Since G_α is 1 : 1, $\mathcal{G} u \in \mathcal{B}(S)$.

<u>Theorem 9</u>. (Dynkin's formula) If $u \in \mathcal{D}(\mathcal{G})$ and σ a Markov time with $E_a(\sigma) < \infty$, then

$$E_a\left(\int_0^\sigma \mathcal{G} u(x_t)\,dt\right) = E_a(u(x_\sigma)) - u(a).$$

Proof as before.

<u>Theorem 10</u>. (Dynkin) If $\mathcal{G} u$ is continuous at a and if $\mathcal{G} u(a) \neq 0$, then

$$\mathcal{G}u(a) = \lim_{U \downarrow a} \frac{E_a(u(x_{\tau_U}))-u(a)}{E_a(\tau_U)}$$

where U denotes a closed neighbourhood of a and τ_U is the leaving time for U, i.e. $\tau_U = \inf\{t : x_t \in (S - U) \cup \infty\}$.

<u>Proof</u>. Since $\mathcal{G}u(a) \neq 0$, we may suppose that $\mathcal{G}u(a) > \alpha > 0$. Let U be a closed neighbourhood of a such that for $b \in U$, $\mathcal{G}u(b) > \alpha/2$. Let $\tau^n = \tau_U \wedge n$; then $E_a(\tau^n) < \infty$ and

$$E_a\left(\int_0^{\tau^n} \mathcal{G}u(x_t)dt\right) = E_a(u(x_{\tau^n})) - u(a).$$

If $t < \tau^n$, $u(x_t) \in U$ and $\mathcal{G}u(x_t) > \alpha/2$. Hence $2\|u\| \geq \frac{\alpha}{2} E_a(\tau^n)$. It follows that $E_a(\tau_U) < \infty$. Therefore

$$E_a\left(\int_0^{\tau_U} \mathcal{G}u(x_t)dt\right) = E_a(u(x_{\tau_U})) - u(a).$$

Since $\mathcal{G}u(a)$ is continuous at a, $\sup_{b \in U} |\mathcal{G}u(a) - \mathcal{G}u(b)| \longrightarrow 0$ as $U \downarrow a$. Therefore

$$\left|\mathcal{G}u(a) - \frac{E_a(u(x_{\tau_U}))-u(a)}{E_a(\tau_U)}\right| = \frac{1}{E_a(\tau_U)}\left| E_a\left(\int_0^{\tau_U} (\mathcal{G}u(x_t) - \mathcal{G}u(a))dt\right)\right|$$

$$\leq \frac{1}{E_a(\tau_U)} E_a(\tau_U) \left|\sup_{b \in U} |\mathcal{G}u(a) - \mathcal{G}u(b)|\right| \longrightarrow 0$$

as $U \downarrow a$

__Theorem 11__. If $u \in \mathcal{D}(\mathcal{G}) = \mathcal{R}$, then given any sequence of Markov times $\{\sigma_n\}$ such that $\sigma_n > 0$, we can find a sequence $\{\tau_n\}$ of Markov times, $0 < \tau_n \leq \sigma_n$ such that

$$\mathcal{G}u(a) = \lim_{n\to\infty} \frac{E_a(u(x_{\tau_n})) - u(a)}{E_a(\tau_n)}$$

__Proof__. Let $\theta_\epsilon(f) = \inf\left\{ t : \frac{1}{t}\left| \int_0^t f(x_s)ds - f(a)\right| > \epsilon \right\}$ for $f \in \mathcal{D}(S)$.

It is easily seen that $\theta_\epsilon(f)$ is a Markov time, and since $P_a(\frac{1}{t}\int_0^t f(x_s)ds \underset{t\to 0}{\longrightarrow} f(a)) = 1$, $P_a(\theta_\epsilon(f) > 0) = 1$. Let now

$$\tau_n = \theta_{1/n}(\mathcal{G}u) \wedge \sigma_n \wedge 1.$$

Then $P_a(\tau_n > 0) = 1$ and $0 < E_a(\tau_n) < 1$. Therefore

$$E_a\left(\int_0^{\tau_n} \mathcal{G}u(x_t)dt\right) = E_a(u(x_{\tau_n})) - u(a).$$

We have

$$\left|\mathcal{G}u(a) - \frac{E_a(u(x_{\tau_n}))-u(a)}{E_a(\tau_n)}\right| \leq \frac{1}{E_a[\tau_n]} E_a\left[\tau_n \frac{1}{\tau_n}\left| \int_0^{\tau_n} (\mathcal{G}u(x_t)-\mathcal{G}u(a))dt\right|\right]$$

$$\leq \frac{1}{E_a(\tau_n)} E_a(\tau_n)\frac{1}{n} \longrightarrow 0 \text{ as } n \longrightarrow \infty.$$

Properties of generator in the restricted sense:

Theorem 12 (Mean value property). Let U be an open subset of S and τ_U the leaving time from $\overline{U}$ and $u \in \mathcal{D}(\mathcal{G})$.

(1) If $u(a) = E_a(u(x_{\tau_U}))$ for every $a \in \overline{U}$, then $\mathcal{G}u(a) = 0$ in U.

(2) Conversely, if $E_a(\tau_U) < \infty$, $\mathcal{G}u(a) = 0$ in U, then

$$u(a) = E_a(u(x_{\tau_U})) \text{ for every } a \in U.$$

Proof. (1) If $u(a) = E_a(u(x_{\tau_U}))$ for every $a \in \overline{U}$, then $u(a) = E_a(u(x_{\tau_U}))$ for every $a \in S$. For, if $a \notin \overline{U}$, $P_a[\tau_U = 0]=1$. It follows that $E_a(u(x_{\tau_U})) = u(a)$. Noting this, let τ be a Markov time $\leq \tau_U$. Then since $\tau_U = \tau + \tau_U(w_\tau^+)$, we have

$$\begin{aligned} u(a) = E_a(u(x_{\tau_U})) &= E_a(u(x_{\tau + \tau_U(w_\tau^+)}(w))) \\ &= E_a(u(x_{\tau_U(w_\tau^+)}(w_\tau^+))) \\ &= E_a(E_{x_\tau}(u(x_{\tau_U}))) = E_a(u(x_\tau)). \end{aligned}$$

Now we can choose a sequence of Markov times $\tau_n = \tau_U \wedge n^{-1} \leq \tau_U$ so that

$$\mathcal{G}u(a) = \lim_{n\to\infty} \frac{E_a(u(x_{\tau_n})) - u(a)}{E_a(\tau_n)} = 0.$$

(2) If $E_a(\tau_U) < \infty$ we have from Dynkin's formula

$$E_a\left(\int_0^{\tau_U} \mathcal{G}u(x_t)dt\right) = E_a(u(x_{\tau_U})) - u(a) ,$$

so that if $\mathcal{G}u(a) = 0$ for $a \in \bar{U}$, $\mathcal{G}u(x_t) = 0$ for $t < \tau_U$ and we get the result.

<u>Theorem 13</u> (Local property). Let $u, v \in \mathcal{D}(\mathcal{G})$ and $u = v$ in a closed neighbourhood U of a. Suppose that there exists a Markov time σ such that $P_a(\sigma > 0) = 1$ and $P_a(x_t$ is continuous for $0 \leq t < \sigma) = 1$. Then

$$\mathcal{G}u(a) = \mathcal{G}v(a).$$

<u>Proof</u>. Let $h = u-v$. Then $h(b) = 0$ for $b \in U$. Let $\tau = \sigma \wedge \tau_U$. Then since x_t is continuous for $0 \leq t \leq \tau$, $x_\tau \in U$ so that $E_a(h(x_\tau)) = 0 = h(a)$. Now

$$\mathcal{G}h(a) = \lim_{n \to \infty} \frac{E_a(h(x_{\tau_n})) - h(a)}{E(\tau_n)} = 0,$$

since τ_n can be chosen so that $\tau_n \leq \sigma \wedge \tau_U$.

Section 3

Multi-dimensional Brownian motion

We have already studied one-dimensional Brownian motion. We shall now define k-dimensional Brownian motion, determine its generator and deduce the main results of Potential Theory using properties of the k-dimensional Brownian motion.

§1. Definition.

We first define k-dimensional Wiener process. Let $x(t,w) = (x_i(t,w),\ i=1,2,\dots,k)$ be a k-dimensional stochastic process on a probability space $\Omega(P)$. $x(t,w)$ is called a k-dimensional Wiener process if 1) its components $x_i(t,w)$ are one-dimensional Wiener processes, and 2) $x_i(t,w)$, $1 \leq i \leq k$, are stochastically independent processes.

It is easy to corstruct a k-dimensional Wiener process $x(t,w)$ on $\Omega(P)$ from a 1-dimensional Wiener process $\xi(t,\lambda)$ on $\Lambda(Q)$. It is sufficient to take $\Omega = \Lambda^k$ and $P =$ the product probability Q^k, and define for $w = (\lambda_1,\dots,\lambda_k)$,

$$x(t,w) = (\xi(t,\lambda_1),\dots,\xi(t,\lambda_k)).$$

We now study the k-dimensional standard Brownian motions. Let $S = R^k$, $W =$ the set of all continuous functions into S and define

$$P_a(B) = P[a+x(\cdot\,,w) \in B].$$

Here $a = (a_1,\dots,a_k)$. It is easily verified that $\mathcal{M} = (S,W,P_a)$ is a Markov process. $\mathcal{M}$ is called the k-dimensional standard Brownian motion. The transition probability of the process is

$$P(t,a,E) = \int_E N_k(t,a,b)db ,$$

where

$$N_k(t,a,b) = N(t,a_1,b_1) \dots N(t,a_k,b_k).$$

Since, for $f \in C(S)$,

$$H_t f(a) = \frac{1}{(2\pi t)^{k/2}} \int e^{-|b|^2/2t} f(a+b)db , \quad |b|^2 = b_1^2 + \dots + b_k^2$$

is also in $C(S)$, $\mathcal{M}$ is a strong Markov process.

Let $\mathcal{O}$ denote the group of congruence (distance-preserving) transformations of R^k. If $O \in \mathcal{O}$, then O induces a transformation, which again we denote by O , of $W \longrightarrow W$ defined by

$$(Ow)(t) = O\,w(t).$$

O carries measurable subsets of W into measurable subsets. For any subset $L \subset W$, we define

$$OL = (Ow : w \in L).$$

The following facts are easily verified:

(0.1) $P(t, Oa, OE) = P(t,a,E)$

(0.2) $P_{Oa}(OB) = P_a(B)$.

If $O \in \mathcal{O}$ is a rotation around a, i.e. if $Oa = a$, $P_a(OB) = P_a(B)$, so that O is a P_a-measure preserving transformation of W onto W.

§2. Generator of the k-dimensional Brownian motion.

Let $\mathcal{D} = \mathcal{D}(R^k)$ be the space of all C^∞-function with compact supports. For $\varphi \in \mathcal{D}$, put

$$\theta(t,a) = H_t \varphi(a),$$

and

$$\psi(a) \equiv \psi(a,\alpha) = G_\alpha \varphi(a) = \int_0^\infty e^{-\alpha t}\, \theta(t,a)dt.$$

Now

$$\begin{aligned}\theta(t,a) &= \int_{R^k} \frac{1}{(2\pi t)^{k/2}} e^{-|b|^2/2t} \varphi(a+b)\,db \\ &= \int_{R^k} \frac{1}{(2\pi)^{k/2}} e^{-|b|^2/2} \varphi(a+b\sqrt{t})\,db,\end{aligned}$$

and a simple calculation gives

$$\frac{\partial \theta}{\partial t} = \tfrac{1}{2}\Delta\theta, \quad \theta(0+,a) = \varphi(a).$$

Taking Laplace transform, the last equation gives

$$(\alpha - \tfrac{1}{2}\Delta)\psi = \varphi.$$

In order to show that ψ is the unique solution of this equation, it is enough to show that if $\psi \in C^2$, $\psi(a) \longrightarrow 0$ as

$|a| \to \infty$ and $(\alpha - \frac{1}{2}\Delta)\psi = 0$, then $\psi \equiv 0$. To prove this, suppose that $\psi(a) > 0$ for some a. Then since $\psi(a) \to 0$ as $|a| \to \infty$, the maximum of $\psi(a)$ is attained at a finite point a_0 and

$$\psi(a_0) = \max \psi(a) > 0.$$

Therefore $$\Delta\psi(a_0) \leq 0,$$

and hence $$(\alpha - \tfrac{1}{2}\Delta)\psi(a_0) > 0.$$

Thus $\psi(a) \leq 0$. Replacing ψ by $-\psi$, we see that $\psi \equiv 0$. This proves our contention.

Now, let $f \in \mathcal{B}(R^k)$. Then

$$u(a) = G_\alpha f(a) = \int G(\alpha, |b-a|) f(b)\,db,$$

where $$G(\alpha, |b-a|) = \int_0^\infty \frac{e^{-\alpha t - |b-a|^2/2t}}{(2\pi t)^{k/2}}\,dt.$$

Note that $G(\alpha, |b-a|)$ is continuous in (a,b). It is immediate that $u \in \mathcal{B}(R^k)$ and we can consider u as a distribution in the Schwartz sense. Then by the definition of the derivative of a distribution, for any $\varphi \in \mathcal{D}$,

$$\begin{aligned}(\alpha - \tfrac{1}{2}\Delta)u(\varphi) &= \int u(a)(\alpha - \tfrac{1}{2}\Delta)\varphi(a)\,da \\ &= \iint G(\alpha, |a-b|) f(b)(\alpha - \tfrac{1}{2}\Delta)\varphi(a)\,da\,db \\ &= \int f(b)\,db \int G(\alpha, |b-a|)(\alpha - \tfrac{1}{2}\Delta)\varphi(a)\,da\end{aligned}$$

$$= \int f(b)\psi(b)db ;$$

where

$$\psi(b) = \int G(\alpha, |a-b|)(\alpha - \tfrac{1}{2}\Delta)\varphi(a)da.$$

If $\theta = (\alpha - \frac{1}{2}\Delta)\varphi$, then $\theta \in \mathcal{D}$ and

$$\psi = G_\alpha \theta ,$$

and from the above we get

$$(\alpha - \tfrac{1}{2}\Delta)\psi = \theta = (\alpha - \tfrac{1}{2}\Delta)\varphi,$$

and hence $$\psi = \varphi.$$

Thus

$$(\alpha - \tfrac{1}{2}\Delta)u(\varphi) = \int f(b)\varphi(b)db,$$

and this means that the distribution $(\alpha - \frac{1}{2}\Delta)u$ is defined by the function f. (Of course, any function equal to f almost everywhere defines the same distribution.)

What we have above also shows that if $u = 0$ then the distribution $(\alpha - \frac{1}{2}\Delta)u = 0$ so that $f = 0$ a.e. Hence

$$\mathcal{N} = \{f : f = 0 \text{ a.e.}\} .$$

Let $\mathcal{R}_+ = \{u : u , \Delta u \in \mathcal{B}(R^k) , \Delta u \text{ is the distribution sense}\}$

$= \{u : u \in \mathcal{B}(R^k)$ and the distribution Δu is defined by a function in $\mathcal{B}(R^k)\}$.

We see from the above that $\mathcal{R} \subset \mathcal{R}_+$. Now suppose $u \in \mathcal{R}_+$. Then $u \in \mathcal{B}(R^k)$ and Δu is defined by a function in $\mathcal{B}(R^k)$. Let $(\alpha - \frac{1}{2}\Delta)u$ be defined by $f \in \mathcal{B}(R^k)$. Put $G_\alpha f = v$ and from the above we see that $(\alpha - \frac{1}{2}\Delta)v$ is defined by f. Hence

$(\alpha u_1 - \frac{1}{2}\Delta u_1) = 0$ where $u_1 = u - v$. We prove that $u_1 = 0$ a.e. Now

$$\int (\alpha - \tfrac{1}{2}\Delta)\varphi(a)u_1(a)da = 0$$

for every $\varphi \in \mathcal{D}(R^k)$, so that

$$\int (\alpha - \tfrac{1}{2}\Delta)\varphi(a+b)u_1(a)da = 0$$

for every $\varphi \in \mathcal{D}(R^k)$. Also

$$\begin{aligned}
&\iint G(\alpha,|b|)|u_1(a)||(\alpha - \tfrac{1}{2}\Delta)\varphi(a+b)|\,da\,db \\
&= \int G(\alpha,|b|)db \int |u_1(a)||(\alpha - \tfrac{1}{2}\Delta)\varphi(a+b)|\,da \\
&= \int G(\alpha,|b|)db \int_{a \in K} |u_1(a-b)||(\alpha - \tfrac{1}{2}\Delta)\varphi(a)|\,da \\
&\leq M \int G(\alpha,|b|)db,
\end{aligned}$$

where K is the compact set outside which φ is zero and

$$M = (\text{diam. } K) \times \text{Sup}|u(a-b)(\alpha - \tfrac{1}{2}\Delta)\varphi(a)|.$$

Therefore,

$$0 = \int G(\alpha,|b|)db \int (\alpha - \tfrac{1}{2}\Delta)\varphi(a+b)u_1(a)da$$

$$= \int u_1(a)da \int G(\alpha, |b|)(\alpha - \tfrac{1}{2}\Delta)\varphi(a+b)db$$

$$= \int u_1(a)da\, \varphi(a).$$

Hence $u_1 = 0$ a.e. Thus

$$\mathcal{R} = \left\{ u : u, \Delta u \in \mathcal{B}(R^k) \right\}$$

and $\mathcal{G}u = \frac{1}{2}\Delta u$ in the distribution sense.

§3. Stochastic solution of the Dirichlet problem.

Let U be a bounded open set and f a function which is bounded and continuous on the boundary ∂U of U. The problem of finding a function $h(a : f,U)$, defined and harmonic in U and such that $h(a : f,U) \longrightarrow f(\xi)$ as $a \longrightarrow \xi \in \partial U$ from within U, is called the Dirichlet problem. $h(a)$, if it exists, is unique and is called the classical solution. The definition of a solution can be generalized, in various ways, so as to include cases in which the classical solution does not exist. The generalized solution will still be harmonic in U, but will tend to the boundary value $f(\xi)$ in a slightly weaker sense.

The stochastic solution, which we shall discuss, gives one way of defining a generalized solution. Let

$$\tau_U = \text{first leaving time from } U = \inf\left\{ t : x_t \notin U \right\}$$

By definition, $u(a) \equiv u(a : f,U) = E_a(f(x_{\tau_U}))$ is the stochastic solution of the Dirichlet problem with boundary value f. We shall see that the stochastic solution is identical with the classical solution, in case the latter exists.

We first establish some results on τ_U.

Theorem 1.

$P_a(\tau_U < \infty) = 1$ if U is a bounded domain.

This is a corollary of the following stronger

Theorem 2.

$E_a[\tau_U] < \infty$, if U is bounded.

Proof. Since $P_a[B+a] = P_o[B]$, we can assume that $a = 0$. Further, since $\tau_U \geq \tau_V$ for $U \supset V$, we can assume that U is the sphere $\Gamma = \{x : |x| < r\}$. Let $u \in \mathcal{D}(\mathcal{G})$ be such that $\mathcal{G}u$ has a version satisfying $\mathcal{G}u(a) > \epsilon_o$ in Γ for some $\epsilon_o > 0$. For example, if $u(a) = - e^{-|a|^2/4r^2}$, then

$$\mathcal{G}u(a) = \tfrac{1}{2}\Delta u(a) = \frac{-1}{2}\left[\frac{k}{2r^2} - \frac{|a|^2}{4r^4}\right]u(a) > 0 \text{ , if } |a| \leq r.$$

Let $\tau_n = \tau_U \wedge n$. Then τ_n is a Markov time, and $E_o(\tau_n) \leq n < \infty$. Therefore, from Dynkin's formula,

$$E_o\left(\int_o^{\tau_n} \mathcal{G}u(x_t)dt\right) = E_o(u(x_{\tau_n})) - u(0)$$

For $0 \leq t \leq \tau_n$, $x_t \in \Gamma$ and $\mathcal{G}u(x_t) \geq \epsilon_o$ and $\epsilon_o E_o(\tau_n) \leq 2\|u\|$

Therefore

$$E_o(\tau_U) = \lim E_o(\tau_n) \leq \frac{2\|u\|}{\epsilon_o} < \infty .$$

Theorem 3.

If U is open and bounded, if f is continuous on ∂U

and if there exists a classical solution $h(a) = h(a : f,U)$, then

$$u(a : f,U) = h(a : f,U).$$

<u>Proof</u>. For any open subset V of U such that $\bar{V} \subset U$, let h_V denote a C^∞ function which vanishes outside U and such that $h_V = h$ or $\bar{V}$. Such a function can easily be constructed. Then $h_V \in \mathcal{D}(\mathcal{G})$. Since V is bounded, $E_a(\tau_V) < \infty$ and Dynkin's formula gives

$$E_a\left(\int_0^{\tau_V} \mathcal{G} h_V(x_t)dt\right) = E_a(h_V(x_{\tau_V})) - h_V(a).$$

For $t < \tau_V$, $x_t \in V$ and $\mathcal{G} h_V(x_t) = \frac{1}{2}\Delta h_V(x_t) = \frac{1}{2}\Delta\, h(x_t) = 0$. If $\tau_V < \infty$, $x_{\tau_V} \in \partial V$ so that $h_V(x_{\tau_V}) = h(x_{\tau_V})$, and since V is bounded, $P_a(\tau_V < \infty) = 1$. Therefore $E_a(h(x_{\tau_V})) = h_V(a)$. Hence, if $a \in \bar{V}$, then $E_a(h(x_{\tau_V})) = h(a)$.

Now let $\{V_n\}$ be an increasing sequence of open subsets of U such that $\bar{V}_n \subset U$ and $V_n \uparrow U$. Then $\tau_U = \lim_{n\to\infty} \tau_{V_n}$. Since $P_a(\tau_U < \infty) = 1$, we have with P_a-measure 1,

$$f(x_{\tau_U}) = \lim_{n\to\infty} h(x_{\tau_{V_n}})$$

$$u(a) = E_a(f(x_{\tau_U})) = \lim_{n\to\infty} E_a(h(x_{\tau_{V_n}})) = h(a)$$

for every $a \in U$. This completes the proof.

A natural question is "When does the classical solution exist?" The simplest case is that of a ball $\Gamma = \Gamma(a_0;r)$. For $a \in \Gamma$, let a' denote the inverse of a with respect to Γ, i.e.

$$a' = a_o + \frac{r^2}{\|a-a_o\|^2}(a-a_o).$$

Let

$$G(b,a) = \frac{1}{|b-a|^{k-2}} - \frac{r^{k-2}}{|a-a_o|^{k-2}} \frac{1}{|b-a'|^{k-2}}, \ k \geqslant 3;$$

$$= \log \frac{1}{|b-a|} - \log \frac{1}{|b-a'|} - \log \frac{r}{|a-a_o|}, \ k = 2,$$

and

$$\pi_\Gamma(a,\xi) = -\frac{1}{k-2} \frac{\partial}{\partial \rho_b} G(b,a)\Big|_{b=\xi} \times r^{k-1} \quad \text{for } k \geqslant 3,$$

$$\pi_\Gamma(a,\xi) = -\frac{\partial}{\partial \rho_b} G(b,a)\Big|_{b=\xi} \times r^{k-1} \quad \text{for } 1 \leqq k \leqq 2,$$

where $\frac{\partial}{\partial \rho_b}$ denotes the derivative in the radial direction of Γ. Then, if f is defined and continuous on the boundary of the ball, and if $\theta(d\xi)$ is the uniform probability distribution (i.e. the normed rotation invariant measure on the boundary of Γ), the classical solution is given by the Poisson integral

$$h(a : f,U) = \int_{\partial\Gamma} \pi_\Gamma(a,\xi) f(\xi)\, \theta(d\xi).$$

The concrete form of $\pi_\Gamma(a,\xi)$ is not of importance to us. The only fact we need is

Theorem 4.

If Γ, V are two concentric balls, with radii $r, \rho\ (r > \rho)$; then

$$c_1 = \min_{a \in \bar{V},\ \xi \in \partial\Gamma} \pi_\Gamma(a,\xi)$$

and $$c_2 = \max_{a\in\overline{\overline{V}},\, \xi\in\partial\Gamma} \pi_\Gamma(a,\xi)$$

depend only on ρ/r and c_1 , $c_2 \to 1$ as $\rho/r \to 0$.

The <u>hitting measure</u> $\pi_U(a,E)$ of E is defined as

$$\pi_U(a,E) = P_a(x_{\tau_U} \in E),\ E\in\mathcal{B}(\partial U).$$

Clearly

$$u(a : f,U) = \int_{\partial U} \pi_U(a,d\xi)f(\xi).$$

We have the following

<u>Theorem 5</u>.

If Γ is a ball, for $a\in\Gamma$ we have

$$\pi_\Gamma(a,d\xi) = \pi_\Gamma(a,\xi)\theta(d\xi)$$

= the harmonic measure on $\partial\Gamma$ with respect to a.

<u>Proof</u>. The proof is immediate since, from the above, for every continuous function f on $\partial\Gamma$,

$$\int_{\partial\Gamma}\pi_\Gamma(a,d\xi)f(\xi) = \int_{\partial\Gamma}\pi_\Gamma(a,\xi)\,\theta(d\xi)f(\xi),$$

and hence the same equation holds for all bounded Borel functions on $\partial\Gamma$.

Using the notation of Theorem 4, we have

<u>Theorem 6</u>.

$$c_1\,\theta(E) \le \pi_\Gamma(a,E) \le c_2\theta(E).$$

We now proceed to prove that if the boundary of a bounded open set U is smooth in a certain sense, then the stochastic solution is also the classical solution.

Definition. Let $\xi \in \partial U$, where U is an open set. If there exists a cone $C \subset U^c$, with vertex at ξ, then ξ is called a Poincaré point for U.

Theorem 7

If ξ is a Poincaré point for a bounded open set U, then for any $\epsilon > 0$ and for any neighbourhood Γ of ξ, there exists a smaller neighbourhood Γ' of ξ such that

$$P_a(x_{\tau_U} \notin \Gamma) < \epsilon$$

for any $a \in \Gamma' \cap U$.

Proof. Let $C \in U^c$ be a cone with vertex at ξ. We can assume that Γ is a ball of radius r such that $C - \Gamma \neq \phi$. Let Γ_n be the ball with the same centre as Γ and radius $r_n = \alpha^n r$, where $\alpha < 1$ is to be chosen subsequently. Let τ_n be the first leaving time from Γ_n. If $x_{\tau_U} \notin \Gamma$, $\tau_\Gamma \leq \tau_U$ and since $P_a(\tau_n \leq \tau_\Gamma) = 1$ for any $a \in \Gamma$ we have $x_{\tau_n} \in U$. Therefore $x_{\tau_n} \notin C$. But $x_{\tau_n} \in \partial\Gamma_n$ so that $x_{\tau_n} \in \partial\Gamma_n - C$. Therefore for any $a \in \Gamma_n$,

$$P_a(x_{\tau_U} \notin \Gamma) \leq P_a(x_{\tau_{n-1}} \in \partial\Gamma_{n-1} - C, \ldots, x_{\tau_1} \in \partial\Gamma_1 - C)$$

$$= P_a(x_{\tau_{n-1}} \in \partial\Gamma_{n-1} - C, \ldots, x_{\tau_2} \in \partial\Gamma_2 - C, x_{\tau_1}(\omega^+_{\tau_2})(\omega^+_\tau) \in \partial\Gamma_1 - C)$$

since $\tau_1 = \tau_2 + \tau_1(w^+_{\tau_2})$. Also since $\tau_i < \tau_2$, for $i > 2$ we have

$$x_{\tau_i} = x(\tau_i(w),w) = x(\tau_i(w),w^-_{\tau_2})$$
$$= x(\tau_i(w^-_{\tau_2}),w^-_{\tau_2}) \in \mathcal{B}_{\tau_2} \subset \mathcal{B}_{\tau_{2+}}.$$

Using the strong Markov property we have

$$P_a(x_{\tau_U} \notin \Gamma) \leq E_a(x_{\tau_{n-1}} \in \partial\Gamma_{n-1} - C, \ldots, x_{\tau_2} \in \partial\Gamma_2 - C : P_{x_{\tau_2}}(x_{\tau_1} \in \partial\Gamma_1 - C)$$

$$\leq c_2\,\theta\, P_a(x_{\tau_{n-1}} \in \partial\Gamma_{n-1} - C, \ldots, x_{\tau_2} \in \partial\Gamma_2 - C),$$

if $a \in \Gamma_n \cap U$, where $\theta = \theta(\partial\Gamma_1 - C) < 1$. Since θ depends only on the solid angle at the vertex ξ, $\theta(\partial\Gamma_1 - C) = \theta(\partial\Gamma_2 - C) = \cdots$ $\ldots = \theta(\partial\Gamma_{n-1} - C)$. We have, repeating the argument,

$$P_a(x_{\tau_U} \notin \Gamma) \leq (c_2\theta)^{n-1}$$

Since $c_2 \longrightarrow 1$ as $\alpha \longrightarrow 0$, we can choose α so small that $c_2\theta < 1$. Now choose n large enough so that $(c_2\theta)^{n-1} < \epsilon$.

<u>Theorem 8</u>.

For any open set U and any bounded Borel function f on ∂U,

$$u(a) = u(a : f, U)$$

is harmonic in U.

<u>Proof</u>. Let $a \in U$ and Γ be a ball with centre at a and contained in U. Then since $\tau_U = \tau_\Gamma + \tau_U(w^+_{\tau_\Gamma})$, we have

$$u(a:f,U) = E_a(f(x_{\tau_U})) = E_a(f(x_{\tau_{U(w^+_{\tau_\Gamma})}}(w^+_{\tau_\Gamma})))$$

$$= E_a(E_{x_{\tau_\Gamma}}(f(x_{\tau_U}))$$

$$= E_a(u(x_{\tau_\Gamma}))$$

$$= \int P_a(x_{\tau_\Gamma} \in d\xi)u(\xi)$$

$$= \int \pi_\Gamma(a,\xi)u(\xi)\theta(d\xi),$$

and the last term is harmonic for $a \in \Gamma$. This proves that u is harmonic in a neighbourhood of every $a \in U$. Hence u is harmonic in U.

Theorem 9.

If U is a bounded open set such that every point of ∂U is a Poincaré point and if f is continuous on ∂U, then the stochastic solution $u = u(a:f,U)$ is also the classical solution.

Proof. By Theorem 8, u is harmonic in U. Let $\xi \in \partial U$. Since f is continuous, we can choose a ball $\Gamma = \Gamma(\xi)$ such that $|f(\eta) - f(\xi)| < \epsilon$ for $\eta \in \Gamma$. By Theorem 7, we can choose Γ' so that

$$P_a(x_{\tau_U} \notin \Gamma) < \epsilon, \; a \in \Gamma'.$$

For $a \in \Gamma'$,

$$|u(a)-f(\xi)| \leq E_a(|f(x_{\tau_U})-f(\xi)|)$$

$$= E_a(|f(x_{\tau_U})-f(\xi)|:x_{\tau_U} \in \Gamma)+E_a(|f(x_{\tau_U})-f(\xi)|:x_{\tau_U} \notin \Gamma)$$

$$\leq \epsilon + 2\,\|f\|\,\epsilon.$$

Remark. When $k = 1$, harmonic functions are linear functions. If (a_1, a_2) is an interval and $f(a_1)$, $f(a_2)$ are given; then

$$h(a:f,(a_1,a_2)) = \frac{a_2 - a}{a_2 - a_1} f(a_1) + \frac{a - a_1}{a_2 - a_1} f(a_2)$$

$$= u(a;f,(a_1,a_2))$$

$$= f(a_1)P_a(x_{\tau(a_1,a_2)} = a_1) + f(a_2)P_a(x_{\tau(a_1,a_2)} = a_2)$$

$$= f(a_1)P_a(\sigma_{a_1} < \sigma_{a_2}) + f(a_2)P_a(\sigma_{a_2} < \sigma_{a_1}),$$

where σ_{a_i}, $i = 1,2$, is the first passage time for a_i, $i=1,2$. Since f is arbitrary,

$$P_a(\sigma_{a_1} < \sigma_{a_2}) = \frac{a_2 - a}{a_2 - a_1},$$

$$P_a(\sigma_{a_2} < \sigma_{a_1}) = \frac{a - a_1}{a_2 - a_1}.$$

We have seen that if U is bounded and open and if every point of ∂U is a Poincaré point, the Dirichlet problem for U has a solution. We now define a generalized solution.

Suppose that U is open and boundod and that f is bounded and continuous on ∂U. Let $\{U_n\} \uparrow U$ be an increasing sequence of open sets with $\bar{U}_n \subset U_{n+1}$ and such that every point of ∂U_n is a Poincaré point. Let F be a continuous extension of f to $\bar{U}$ and $F_n = F / \partial U_n$. Denote the classical solution for U_n with boundary values F_n by $h(a;F_n,U_n)$. Then, $\lim_{n\to\infty} h(a;F_n,U_n)$ is, by definition, the generalized solution (in the Wiener sense) of the Dirichlet problem with boundary values f. We have of course to show that the limit exists and is independent of the choice of U_n and of F.

Theorem 10.

For a bounded open set U, $u(a;f,U)$ is the generalized solution.

Proof. We have only to show that $h(a:F_n,U_n) \longrightarrow u(a:f,U)$. In fact since $\tau_{U_n} \uparrow \tau_U < \infty$ with probability 1,

$$h(a:F_n,U_n) = u(a:F_n,U_n) = E_a(F(x_{\tau_{U_n}}))$$

$$\longrightarrow E_a(F(x_{\tau_U}))$$

$$= E_a(f(x_{\tau_U})) = u(a:f,U) .$$

Remark. $u(a) = u(a:f,U)$ does not always satisfy the boundary condition $\lim_{a\in U, a\to\xi} u(a) = f(\xi)$ for $\xi \in \partial U$. In § 7, we shall discuss these boundary conditions.

§4. Recurrence.

Definition. A Markov process M is called recurrent if

$$P_a(x_t \in U \text{ for some } t) \equiv P_a(\sigma_U < \infty) \doteq 1$$

for any $a \in S$ and any open U; otherwise it is called non-recurrent.

We shall now show that the standard Brownian motion is recurrent for $k \leq 2$ and is non-recurrent for $k \geq 3$.

Theorem 1.

Let Γ_1, Γ_2 be two balls with centres a_o and radii r_1, $r_2 (r_2 > r_1)$. If $\sigma_1 = \sigma_{\partial\Gamma_1}$, $\sigma_2 = \sigma_{\partial\Gamma_2}$ are the first passage times for $\partial\Gamma_1$ and $\partial\Gamma_2$, then for $a \in \Gamma_2 - \overline{\Gamma}_1$,

$$P_a(\sigma_1 < \sigma_2) = \begin{cases} \dfrac{r^{-k+2} - r_2^{-k+2}}{r_1^{-k+2} - r_2^{-k+2}}, & k \geqslant 3 ; \\[2ex] \dfrac{\log 1/r - \log 1/r_2}{\log 1/r_1 - \log 1/r_2}, & k = 2 ; \\[2ex] \dfrac{r_2 - r}{r_2 - r_1}, & k = 1 ; \end{cases}$$

where $r = |a-a_o|$.

Proof. In fact, if $U = \Gamma_2 - \overline{\Gamma}_1$, $\partial U = \partial\Gamma_1 \cup \partial\Gamma_2$, and the function f which is 1 on $\partial\Gamma_1$ and 0 on $\partial\Gamma_2$ is continuous on ∂U. Since every point in ∂U is a Poincaré point, the classical solution $h(a; f,U) = u(a;f,U)$ exists and

$$p(a) \equiv P_a(\sigma_1 < \sigma_2) = P_a(x_{\tau_U} \in \partial\Gamma_1) = u(a;f,U).$$

The function given in the statement of the theorem is harmonic in U and takes the boundary value f. Since such a function is unique, we get the result.

Theorem 2.

Let $\Gamma = \Gamma(a_o, r)$ be a ball with centre a_o and radius r and let σ_Γ be the first passage time for Γ. For $a \notin \Gamma$ and

$\rho = |a-a_o|$,

$$P_a(\sigma_\Gamma < \infty) = \begin{cases} (r/\rho)^{k-2} & , k \geq 3 \\ 1 & , k \leq 2. \end{cases}$$

Therefore k-dimensional Brownian motion is recurrent or not according as $k \leq 2$ or $k \geq 3$.

Proof. Observe that $\sigma_\Gamma = \sigma_{\partial\Gamma}$ for any path whose starting point is not in Γ. Let $\Gamma' = \Gamma'(a_o, r')$ and $\sigma' = \sigma_{\partial\Gamma'}$. If $t < \sigma_\infty(w)$, then since $w(t)$ is continuous, $F_t = \{x_s : 0 \leq s \leq t\}$ is a compact set and hence we can find r' such that $\Gamma' \supset F_t$. Then $\sigma'(w) \geq t$. It follows that

$$\lim_{r' \to \infty} \sigma' = \infty .$$

Therefore

$$P_a(\sigma_\Gamma < \infty) = P_a(\sigma_\Gamma < \lim_{r' \to \infty} \sigma') = \lim_{r' \to \infty} P_a'(\sigma_\Gamma < \sigma').$$

Now take $r_2 = r'$ and $\sigma_2 = \sigma'$ in Theorem 1 and we get the result.

Theorem 3.

If $k \geq 3$, $P_a(|x_t| \to \infty$, as $t \to \infty) = 1$.

If $k \leq 2$, $P_a(w : (x_s, s \geq t$, is dense in R^k for all $t)) = 1$.

Proof. Case $k \geq 3$. We can, without loss of generality, assume that $a = 0$. Let $\Gamma_n = \Gamma(o,n)$ and $\sigma_n = \sigma_{\partial\Gamma_n}$. For any path w , $|x_t| \to \infty$ if and only if for every given n we can find s such that the image of $[o,\infty]$ by w_s^+ is contained in Γ_n^c. Therefore $|x_t| \not\to \infty$ if and only if we can find n such that for every $s \geq 0$, the image of $[o,\infty)$ by w_s^+ has a non-empty intersection with Γ_n and therefore if $w_s^+(0) \notin \Gamma_n$, then $\sigma_n(w_s^+) < \infty$. Therefore

$$P_o[|x_t| \not\to \infty] =$$

$$= P_o[\exists n \text{ such that for every } s \geq 0 \text{ with } w_s^+(0) \notin \Gamma_n, \sigma_n(w_s^+) < \infty]$$

$$\leq \sum_n P_o[\text{for every } s \geq 0 \text{ with } w_s^+(0) \notin \Gamma_n, \sigma_n(w_s^+) < \infty]$$

$$\leq \sum_n P_o[\text{for every } m > n, \sigma_n(w^+_{\sigma_m}) < \infty].$$

Now

$$P_o(\text{for every } m, \sigma_n(w^+_{\sigma_m}) < \infty) \leq P_o(\sigma_n(w^+_{\sigma_m}) < \infty \text{ for some } m)$$

$$= E_o(P_{x_{\sigma_m}}(\sigma_n < \infty)$$

$$= (\tfrac{n}{m})^{k-2} \longrightarrow 0, \text{ as } m \longrightarrow \infty.$$

Case $k \leq 2$. Let Γ be any ball and σ_Γ = the first passage time for Γ. We have

$$P_a(\sigma_\Gamma < \infty) = 1$$

for every a, so that for any t,

$$P_a(\sigma_\Gamma(w_t^+) < \infty) = E_a(P_{x_t}(\sigma_\Gamma < \infty)) = 1.$$

Now

$$P_a(\text{for every } t\ ,\ \sigma_{\Gamma}(w_t^+) < \infty) =$$

$$= P_a(\text{for every } n\ ,\ \sigma_{\Gamma}(w_n^+) < \infty)$$

$$= 1.$$

Let $\Gamma_1, \Gamma_2, \ldots$ be a complete fundamental system of neighbourhoods. Then

$$P_a(\text{for every } n\ ,\ \text{for every } t\ ,\ \sigma_{\Gamma_n}(w_t^+) < \infty) = 1,$$

i.e. $$P_a((x_s(w) : s \geq t \ \text{ is dense in } R^k)) = 1.$$

§5. Green function.

Case $k \geq 3$.

Definition. Let U be a bounded open set. Then

$$G_U(a,b) = \frac{1}{|a-b|^{k-2}} - \int_{\partial U} \frac{\pi_U(a,d\xi)}{|\xi-b|^{k-2}}$$

is called the Green function for U, where $\pi_U(a,d\xi)$ is the harmonic measure on ∂U with respect to a. This is the potential at b due to a unit charge at a and the induced charge on ∂U.

As the limiting case, when $U \longrightarrow R^k$, we can define the Green function (relative to the whole space R^k by

$$G(a,b) = \frac{1}{|a-b|^{k-2}}.$$

Theorem 1.

If f is bounded, Borel and has compact support, then $E_a(\int_0^\infty f(x_t)dt) < \infty$ and

$$E_a(\int_0^\infty f(x_t)dt) = \frac{2}{K}\int \frac{f(b)db}{|b-a|^{k-2}}, \text{ where } K := 4\pi^{k/2}/\Gamma(k/2-1)$$

Proof. It is enough to prove the theorem for $f \geq 0$. We have

$$\begin{aligned} E_a(\int_0^\infty f(x_t)dt) &= \int_0^\infty E_a(f(x_t))dt \\ &= \int_0^\infty dt \int_{R^k} \frac{1}{(2\pi t)^{k/2}} e^{-\frac{|b-a|^2}{2t}} f(b)db \\ &= \int_{R^k} f(b)db \int_0^\infty \frac{1}{(2\pi t)^{k/2}} e^{-\frac{|b-a|^2}{2t}} dt \\ &= \int_{R^k} \frac{f(b)db}{|b-a|^{k-2}} \frac{\Gamma(k/2-1)}{2\pi^{k/2}} \\ &= \frac{2}{K}\int_{R^k} \frac{f(b)db}{|b-a|^{k-2}} \end{aligned}$$

$$< \infty ,$$

because, if Γ is a ball containing the support of f,

$$\int_\Gamma \frac{f(b)db}{|b-a|^{k-2}} \leq \|f\| \int_\Gamma \frac{db}{|b-a|^{k-2}} < \infty .$$

Theorem 2.

For f as in Theorem 1, $v(a) := E_a\left(\int_0^\infty f(x_t)dt\right) \in \mathcal{D}(\mathcal{G})$,

$\frac{1}{2}\Delta v = -f$ a.e., and $v(a) \longrightarrow 0$ as $|a| \longrightarrow \infty$.

Therefore, if $u(a) = \int G(a,b)f(b)db$,

$$\Delta u = -K f \text{ a.e. (Poisson's equation)}$$

and $u(a) \longrightarrow 0$ as $|a| \longrightarrow \infty$.

Proof. By Theorem 1, $v(a)$ is bounded and Borel. If

$$G_\epsilon f(a) = E_a\left(\int_0^\infty e^{-\epsilon t} f(x_t)dt\right),$$

we have

$$v(a) = \lim_{\epsilon \longrightarrow 0} G_\epsilon f(a)$$

and the resolvent equation gives

$$G_\alpha f - G_\epsilon f + (\alpha - \epsilon)G_\alpha G_\epsilon f = 0.$$

Letting $\epsilon \longrightarrow 0$,

$$G_\alpha f - v + \alpha G_\alpha v = 0,$$

or

$$v = G_\alpha (f + \alpha v) \in \mathcal{D}(\mathcal{G}).$$

Also, since $\mathcal{G} v = \alpha v - G_\alpha^{-1} v = \alpha v - f - \alpha v = -f$, a.e.,

$$\frac{1}{2}\Delta v = -f \text{ a.e.}$$

Definition. Let A be a bounded subset of R^k. Then

$$S(A,w) = \text{the Lebesgue measure of } \left\{ t : x_t(w) \in A \right\}$$

is called the sojourn (visiting) time for the set A.

From Theorem 1, we have

Theorem 3.

$$\frac{E_a(S(db))}{db} = \frac{2}{K} G(a,b).$$

Let now U be a bounded open set and $f \in \mathcal{B}(U)$. Let

$$v_U(a) = v_U(a;f,U) = E_a\left(\int_0^{\tau_U} f(x_t)dt \right),$$

τ_U being the first leaving time from U.

Theorem 4.

$$v_U(a) = \frac{2}{K} \int_U G_U(a,b)f(b)db.$$

Proof. Extend f by putting $f = 0$ in U^c. Then

$$v_o(a) = E_a\left(\int_o^\infty f(x_t)dt \right) = \frac{2}{K} \int_U \frac{f(b)db}{|b-a|^{k-2}},$$

by Theorem 1. Also

$$\begin{aligned} v_o(a) &= E_a\left(\int_o^{\tau_U} f(x_t)dt \right) + E_a\left(\int_{\tau_U}^\infty f(x_t)dt \right) \\ &= v_U(a) + E_a\left(\int_o^\infty f(x_t(w^+_{\tau_U}))dt \right) \\ &= v_U(a) + E_a\left(E_{x_{\tau_U}}\left(\int_o^\infty f(x_t)dt \right) \right) \\ &= v_U(a) + E_a(v_o(x_{\tau_U})). \end{aligned}$$

$$= v_U(a) + \int_{\partial U} \Pi_U(a,d\xi) v_o(\xi)$$

$$= v_U(a) + \frac{2}{K} \int f(b)db \int_{\partial U} \frac{\Pi_U(a,d\ )}{|b-\xi|^{k-2}}$$

This gives the result.

<u>Theorem 5</u>. $v_U(a)$ satisfies

$$\tfrac{1}{2} \Delta v_U = - f \ , \text{ a.e.,}$$

and $v_U(a) \longrightarrow 0$ as $a \longrightarrow \xi$, ξ being a regular point of ∂U.

<u>Proof</u>. $v_U(a) = v_o(a) - E_a(v_o(x_{\tau_U}))$. Since $E_a(v_o(x_{\tau_U}))$ is harmonic in U. and $\frac{1}{2}\Delta v_o(a) = - f$ a.e., we have

$$\tfrac{1}{2}\Delta v_U(a) = - f \ , \text{ a.e.}$$

Further if $\xi \in \partial U$ is regular, $E_a(v_o(x_{\tau_U})) \longrightarrow v_o(\xi)$ as $a \longrightarrow \xi$, and since $v_o(a)$ is continuous by Theorem 1, $v_o(a) \longrightarrow v_o(\xi)$ as $a \longrightarrow \xi$. The result follows.

<u>Theorem 6</u>.

Let $S(A/U,w)$ = the Lebesgue measure of $\{t : x_t \in A, \ t < \tau_U\}$. Then

$$\frac{E_a(S(db/U))}{db} = \frac{2}{K} G_U(a,b).$$

As an example we compute $v_U(a)$ for U = the open cube $(0,1)^3$, k = 3. Since every boundary point of the unit cube is regular (in fact every point is a Poincaré point), $v_U = 0$ as ∂ U. Therefore $v = v_U(a)$ is the solution of

$$\tfrac{1}{2}\Delta v = -f \text{ and } v = 0 \text{ on } \partial U.$$

Since $v = 0$ on ∂U, we can put

$$v(x,y,z) = \sum_{\ell+m+n>0} a_{\ell mn} \sin \ell\pi x \sin m\pi y \sin n\pi z .$$

Then

$$\tfrac{1}{2}\Delta v = -\frac{\pi^2}{2} \sum_{\ell+m+n>0} (\ell^2+m^2+n^2) a_{\ell mn} \sin \ell\pi x \sin m\pi y \sin n\pi z$$

If

$$f(x,y,z) = \sum b_{\ell mn} \sin \ell\pi x \sin m\pi y \sin n\pi z,$$

we have therefore

$$a_{\ell mn} = \frac{2b_{\ell mn}}{\pi^2(\ell^2+m^2+n^2)}$$

$$= \frac{16}{\pi^2(\ell^2+m^2+n^2)} \int_0^1\int_0^1\int_0^1 f(\xi,\eta,\zeta) \sin \ell\pi\xi \sin m\pi\eta \sin n\pi\zeta \, d\xi\, d\eta\, d\zeta$$

This gives

$$v(x,y,z) = \iiint f(\xi,\eta,\zeta)\frac{16}{\pi^2}\sum \frac{\sin \ell\pi\xi \sin \ell\pi x \sin m\pi\eta \sin m\pi y \sin n\pi\zeta \sin n\pi z}{\ell^2+m^2+n^2} \, d\xi\, d\eta\, d\zeta .$$

Hence

$$G_U(x,y,z;\xi,\eta,\zeta) = \frac{32}{\pi}\sum \frac{\sin \ell\pi\xi \sin \ell\pi x \sin m\pi\eta \sin m\pi y \sin n\pi\zeta \sin n\pi z}{\ell^2+m^2+n^2}$$

in the distribution sense.

Case $k \leq 2$.

We cannot apply the preceding method to discuss the Green function for $k \leq 2$ because $E_a(\int_0^\infty f(x_t)dt)$ may be infinite even if f has compact support. We therefore follow a different method.

Let $\Gamma = \Gamma(o,r)$ be a ball. If $u \in C^\infty(R^2)$, [i.e. compact support and C^∞] then $u \in \mathcal{D}(\mathcal{G})$ and Dynkin's formula gives

$$E_a\left(\int_0^{\tau_\Gamma} \tfrac{1}{2}\Delta u(x_t)dt\right) = E_a(u(x_{\tau_\Gamma})) - u(a)$$

$$= \int_{\partial\Gamma} \pi_\Gamma(a,\xi)u(\xi)\theta(d\xi) - u(a), \quad \pi_\Gamma(a,\xi) = \frac{r^2 - a^2}{|a-\xi|^2},$$

$$= -\int_{\partial\Gamma} \left.\frac{\partial G_\Gamma(a,b)}{\partial n}\right|_{b=\xi\in\partial\Gamma} r u(\xi)\theta(d\xi) - u(a), \quad G_\Gamma(a,b) = \log\frac{|a\bar{b}-r^2|}{|a-b|}$$

$$= \int_\Gamma \frac{1}{2\pi} G_\Gamma(a,b)\, \tfrac{1}{2}\Delta u(b) 2\, db.$$

If $\varphi \in C^\infty(R^2)$ and $v(a) = \frac{1}{\pi}\int_\Gamma G_\Gamma(a,b)\varphi(b)db$, then $\frac{1}{2}\Delta v = \varphi$. Therefore we have for any $\varphi \in C^\infty(R^2)$,

$$E_a\left(\int_0^{\tau_\Gamma} \varphi(x_t)dt\right) = \frac{1}{\pi}\int_\Gamma G_\Gamma(a,b)\varphi(b)db.$$

It follows that the same equation holds for any $f \in \mathcal{B}(R^2)$, i.e.

$$E_a\left(\int_0^{\tau_\Gamma} f(x_t)dt\right) = \frac{1}{\pi}\int_\Gamma G_\Gamma(a,b)f(b)db.$$

Now let U be a bounded domain, $\bar{U} \subset \Gamma$, a ball. Then

$$E_a\Big(\int_0^{\tau_\Gamma} f(x_t)dt\Big) = E_a\Big(\int_0^{\tau_U} f(x_t)dt\Big) + E_a\Big(\int_0^{\tau_\Gamma(w^+_{\tau_U})} f(x_t(w^+_{\tau_U}))dt\Big)$$

$$= E_a\Big(\int_0^{\tau_U} f(x_t)dt\Big) + E_a\Big(E_{x_{\tau_U}}\Big(\int_0^{\tau_\Gamma} f(x_t)dt\Big)\Big) ,$$

so that

$$E_a\Big(\int_0^{\tau_U} f(x_t)dt\Big) = \frac{1}{\pi}\int_\Gamma G_\Gamma(a,b)f(b)db - \int_{\partial U}\pi_U(a,d\xi)E_\xi\Big(\int_0^{\tau_\Gamma} f(x_t)dt\Big)$$

$$= \frac{1}{\pi}\int_\Gamma G_\Gamma(a,b)f(b)db - \frac{1}{\pi}\int_{\partial U}\pi_U(a,d\xi)\int_\Gamma G_\Gamma(\xi,b)f(b)db$$

$$= \frac{1}{\pi}\int_\Gamma G_U(a,b)f(b)db ,$$

where

$$G_U(a,b) = G_\Gamma(a,b) - \int_{\partial U}\pi_U(a,d\xi)G_\Gamma(\xi,b)$$

$$= \log\frac{1}{|a-b|} - \int_{\partial U}\pi_U(a,d\xi)\log\frac{1}{|\xi-b|} - \log\frac{1}{|a\bar{b}-r^2|} + \int_{\partial U}\pi_U(a,d\xi)\log\frac{1}{|\xi\bar{b}-r^2|}$$

Since $b \in U \subset \bar{U} \subset \Gamma$, $|a\bar{b}| < r^2$ for $a \in U$ and $\log|a\bar{b} - r^2|$ is harmonic for $a \in U$, with boundary values $\log|\xi\bar{b} - r^2|$. Hence if every point of ∂U is regular,

$$\log|a\bar{b}-r^2| = \int_{\partial U}\pi_U(a,d\xi)\log|\xi\bar{b}-r^2|.$$

Thus we have

$$G_U(a,b) = \log\frac{1}{|a-b|} - \int_{\partial U}\pi_U(a,d\xi)\log\frac{1}{|\xi-b|}.$$

Theorem 1.

If U is a bounded open set such that every point of ∂U

is regular and if $u(a) = E_a(\int_0^{\tau_U} f(x_t)dt)$, then

$$\tfrac{1}{2}\Delta u = f \text{ and } u(a) \longrightarrow 0 \text{ as } a \longrightarrow \xi \in \partial U.$$

Proof. In fact

$$u(a) = E_a(\int_0^{\tau_U} f(x_t)dt) = \frac{1}{\pi}\int_U G_U(a,b)f(b)db$$

and the theorem follows from the definition of $G_U(a,b)$.

Theorem 2.

$$\frac{E_a(S(db/U))}{db} = \frac{1}{\pi} G_U(a,b).$$

If $k=1$, we can proceed directly. Suppose that $U = (\alpha,\beta)$.

Then

$$\begin{aligned} E_a(\int_0^{\tau(\alpha,\beta)} \tfrac{1}{2} u''(x_t)dt) &= E_a(u(x_{\tau(\alpha,\beta)})) - u(a) \\ &= \frac{\beta-a}{\beta-\alpha} u(\alpha) + \frac{a-\alpha}{\beta-\alpha} u(\beta) - u(a) \\ &= \int_\alpha^\beta G_{(\alpha,\beta)}(a,b)\ \tfrac{1}{2} u''(b)2db, \end{aligned}$$

where

$$G_{(\alpha,\beta)}(x,y) = G_{(\alpha,\beta)}(y,x) = \frac{(\beta-y)(x-\alpha)}{\beta-\alpha},\quad \alpha \le x \le y \le \beta .$$

Therefore we have

Theorem 3

$$E_a(\int_0^{\tau(\alpha,\beta)} f(x_t)dt) = \int_\alpha^\beta G_{(\alpha,\beta)}(a,b)f(b)2db$$

Theorem 4

$$\frac{E_a(S(db/(\alpha,\beta)))}{db} = 2\, G_{(\alpha,\beta)}(a,b).$$

§6. Hitting probability ($k \geq 3$).

We have already discussed the hitting probability for spheres. Here we shall discuss it for more general sets, especially compact sets.

Absolute hitting probability ($k \geqslant 3$).

For simplicity we consider the case $k = 3$.

Let F be a compact set and $\sigma_F = \inf\{t : t > 0$ and $x_t \in F\}$. Put $p_F(a) = P_a(\sigma_F < \infty) = P_a(x_t \in F$ for some $t > 0)$; $p_F(a)$ is called the absolute hitting probability for F with respect to a.

Lemma. Let $\Gamma = \Gamma(a,r)$ and $\tau_r = \tau_\Gamma =$ the first leaving time for Γ. Then $P_a(\tau_r \longrightarrow 0$ as $r \longrightarrow 0) = 1$.

Proof. Clearly τ_r decreases as r decreases. If $\tau = \lim_{r \to 0} \tau_r$, we have only to show that $P_a(\tau > 0) = 0$. Now $P_a(\tau > t) \leq$
$\leq P_a(\tau_r > t) \leq P_a(x_t \in \Gamma) = (2\pi t)^{-3/2} \int_\Gamma \exp(-\|x-a\|^2 \times \frac{1}{2t}) dx \longrightarrow 0$ as $r \longrightarrow 0$.

Theorem 1.

$p_F(a)$ is expressible as a potential induced by a bounded measure μ_F i.e. $p_F(a) = \int \frac{\mu_F(db)}{|b-a|}$, where μ_F is concentrated on ∂F. Further μ_F is uniquely determined by F.

Proof. Firstly we show that $p_F(a)$ is harmonic in $F^c \cup F^0$. Since $p_F(a) \equiv 1$ in F^0, we have only to show that it is harmonic in F^c. Let Γ be ball such that $\bar{\Gamma} \subset F^c$. If τ_Γ is the first leaving time for Γ, $\tau_\Gamma < \sigma_F$ and $p_F(a) = P_a(\tau_\Gamma < \sigma_F < \infty) = P_a(\sigma_F(w^+_{\tau_\Gamma}) < \infty) =$
$= E_a(P_{x_{\tau_\Gamma}}(\sigma_F < \infty)) = E_a(p_F(x_{\tau_\Gamma})) = \int_{\partial\Gamma} \pi_\Gamma(a, d\xi) p_F(\xi) =$
$= \int_{\partial\Gamma} \pi_\Gamma(a, \xi) p_F(\xi) \theta(d\xi)$, showing that $p_F(a)$ is harmonic for $a \in \Gamma$.

Let Γ be a ball and $a \in \Gamma$. Then

$$p_F(a) = P_a(\sigma_F < \infty) \geq p_a(\sigma_F(w^+_{\tau_\Gamma}) < \infty) = \int_{\partial\Gamma} \pi_\Gamma(a,\xi) p_F(\xi)\, \theta(d\xi).$$

This shows that $p_F(a)$ is super harmonic in the wide sense (i.e. its value at the centre of a ball is not less than the average value on the boundary).

Finally we show that $p_F(a)$ is lower semi-continuous. It is enough to show this for $a \in \partial F$. Let $a_0 \in \partial F$, and $\Gamma(a_0, r) = \Gamma_r$, $\tau_r = \tau_{\Gamma_r}$. Then $p_F(a_0) = P_{a_0}(x_t \in F$ for some $t > \tau_r) + \eta_r$, $\eta_r \longrightarrow 0$ (from the lemma) $= \int_{\partial\Gamma_r} p_F(\xi)\, \theta(d\xi) + \eta_r$. On the other hand

$$p_F(a) \geq \int_{\partial\Gamma_r} \pi_\Gamma(a,\xi) p_F(\xi)\, \theta(d\xi)$$

so that

$$\varliminf_{a \to a_0} p_F(a) \geq \int_{\partial\Gamma_r} \varliminf_{a \to a_0} \pi_\Gamma(a,\xi) p_F(\xi) \theta(d\xi) = \int_{\partial\Gamma_r} p_F(\xi)\, \theta(d\xi) = p_F(a_0) - \eta_r .$$

Now letting $r \longrightarrow 0$, $\varliminf_{a \to a_0} p_F(a) \geq p_F(a_0)$, showing that $p_F(a)$ is lower semi-continuous. Now if Γ is a ball containing F, σ_Γ, the first passage time for Γ, then we have seen that $P_a(\sigma_\Gamma < \infty) = r\rho^{-1}$, $\rho = |a|$. Therefore $P_a(\sigma_\Gamma < \infty) \longrightarrow 0$ as $a \longrightarrow \infty$ and since $P_a(\sigma_F < \infty) \leq P_a(\sigma_\Gamma < \infty)$, $P_a(\sigma_F < \infty) \longrightarrow 0$ as $a \longrightarrow \infty$. Since $p_F(a)$ is super harmonic in R^3, from the Riesz representation theorem there exists a unique bounded measure μ_F with $p_F(a) = \int \frac{\mu_F(db)}{|b-a|} + H(a)$ where $H(a)$ is harmonic in R^3. But $p_F(a) \longrightarrow 0$

$|a| \to \infty$ and also $\int \frac{\mu_F(db)}{|b-a|} \to 0$ as $|a| \to \infty$ since μ_F is a bounded measure. It follows that $H(a) \to 0$ as $|a| \to \infty$ i.e. $H(a) \equiv 0$. Therefore $p_F(a) = \int \frac{\mu_F(db)}{|b-a|}$. Since μ_F is concentrated in the set where p_F is not harmonic, μ_F is concentrated in ∂F. This proves the theorem completely.

Theorem 2.

If $u(a)$ is any potential induced by a measure ν which is concentrated in F and if $u(a) \leq 1$, then

$$u(a) \leq p_F(a) \text{ and } \nu(F) \leq \mu_F(F)(=\mu_F(\partial F)).$$

Proof. We have $u(a) = \int_F \frac{\nu(db)}{|a-b|}$. Since F is compact, for fixed a we can find n such that $|a-b| \leq n$. It follows that $\nu(F) < \infty$ and therefore $u(a)$ is harmonic in F^c. Let $G_n \uparrow F^c$ be a sequence of bounded open sets such that $\bar{G}_n \subset G_{n+1}$. Let $\tau_n = \tau_{G_n}$ the first leaving time from G_n. If we put $f = u|G_n$ then u is the classical solution with boundary values f. Therefore, for every $a \in G_n$,

$$u(a) = E_a(f(x_{\tau_n})) = E_a(u(x_{\tau_n})) = E_a(u(x_{\tau_n}) : \sigma_F = \infty) +$$

$$+ E_a(u(x_{\tau_n}) : \sigma_F < \infty).$$

Now $\tau_n \uparrow \sigma_F$. If $\sigma_F = \infty$, $\tau_n \uparrow \infty$ and $x_{\tau_n} \uparrow \infty$ with probability 1; and by the formula for u, $u(x_{\tau_n}) \to 0$. Since $u(a) \leq 1$ we have therefore

$$u(a) \leq E_a(\sigma_F < \infty) = p_F(a).$$

If $a \in F^{\circ}$, $p_F(a) = 1$ and $u(a) \leq 1 = p_F(a)$.

Let now $a \in \partial F$ and $\Gamma = \Gamma(a,r)$ and τ_r the first leaving time for Γ. Then

$$p_F(a) \geqslant P_a(x_t \in F \text{ for some } t \geqslant \tau_r) = P_a(x_t(w^+_{\tau_r}) \in F \text{ for some } t \geqslant 0)$$

$$= E_a\left[P_{x_{\tau_r}}(x_t \in F \text{ for some } t \geqslant 0)\right]$$

$$= E_a\left[P_{x_{\tau_r}}(x_t \in F \text{ for some } t \geqslant 0) : x_{\tau_r} \in F^c)\right] + E_a\left[P_{x_{\tau_r}}(x_t \in F \text{ for some } t \geqslant 0) : x_{\tau_r} \in F)\right.$$

$$\geqslant E_a\left[x_{\tau_r} \in F^c : u(x_{\tau_r})\right] + E_a\left[x_{\tau_r} \in F : 1\right] \geqslant E_a\left[u(x_{\tau_r})\right]$$

since $P_a(x_t \in F \text{ for some } t \geqslant 0) = 1$ for $a \in F$. Letting $r \longrightarrow 0$ we get

$$p_F(a) \geqslant \underline{\lim}_{r \to 0} E_a(u(x_{\tau_r})) \geqslant E_a(\underline{\lim}_{r \to 0} u(x_{\tau_r})) \geqslant u(a)$$

since $u(a)$ is lower semi-continuous. It remains to prove that $\nu(F) \leq \mu_F(F)$.

Let E be a compact set with $E \supset E^{\circ} \supset F$ and consider $p_E(a)$. Then $p_E(a) = \int \frac{\mu_E(db)}{|a-b|}$ and $p_E(a) = 1$ for $a \in E^{\circ} \supset F$.

Since

$$\int \frac{\mu_F(db)}{|a-b|} \geqslant \int \frac{\nu(db)}{|b-a|}$$

we have

$$\iint \frac{\mu_F(db)}{|b-a|} \mu_E(da) \geqslant \iint \frac{\nu(db)}{|b-a|} \mu_E(da)$$

i.e. $$\int_F \mu_F(db) \geqslant \int_F \nu(db).$$

An alternative proof of the last fact is the following. Since $p_F(a) \geqslant u(a)$,

$$\int_F |a| \frac{\mu_F(db)}{|b-a|} \geqslant \int_F |a| \frac{\nu(db)}{|b-a|}$$

Letting $a \longrightarrow \infty$, we get the result.

From the above theorem we have

Theorem 3. $C(F) = \mu_F(\partial F)$ is the maximal total charge for those charge distributions which induce potentials ≤ 1.

Theorem 4 (Kakutani). $C(F) > 0$ if and only if $p_F(a) > 0$ i.e.

$$P_a(x_t \in F \text{ for some } t > 0) > 0.$$

$C(F)$ is called the capacity of F.

Theorem 5. $\dfrac{C(F)}{\max\limits_{b\in F}|a-b|} \leq p_F(a) \leq \dfrac{C(F)}{\min\limits_{b\in F}|b-a|}$ and $C(F) = \lim\limits_{|a|\to\infty} |a| p_F(a)$.

We shall now prove the subadditivity of $p_F(a)$ and $C(F)$ following Hunt. This means that $p_F(a)$ and $C(F)$ are both strong capacities in the sense of Choquet.

Theorem 6.

$p_F(a)$ and $C(F)$ are subadditive in the following sense.

$$\varphi(F_1 \cap \ldots \cap F_n) \leq \sum_i \varphi(F_i) - \sum_{i<j} \varphi(F_i \cup F_j) + \sum_{i<j<k} \varphi(F_i \cup F_j \cup F_k) \ldots$$

where $\varphi(F)$ denotes either of $p_F(a)$ and $C(F)$.

Proof. Put $F^* = \{w : \sigma_F(w) < \infty\}$. Then $(F_1 \cup \ldots \cup F_n)^* = F_1^* \cup \ldots \cup F_n^*$, $(F_1 \cap \ldots \cap F_n)^* \subset F_1^* \cap \ldots \cap F_n^*$ and $p_F(a) = P_a(F^*)$. Using the dual inclusion-exclusion formula of Hunt, we have

$$p_{F_1 \cap \ldots \cap F_n}(a) = P_a\left[(F_1 \cap \ldots \cap F_n)^*\right] \leq P_a\left[F_1^* \cap \ldots \cap F_n^*\right]$$

$$= \sum_i P_a(F_i^*) - \sum_{i<j} P_a(F_i^* \cup F_j^*) + \ldots\ldots$$

$$= \sum_i p_{F_i}(a) - \sum_{i<j} p_{F_i \cup F_j}(a) + \sum_{i<j<k} p_{F_i \cup F_j \cup F_k}(a) \ldots.$$

Multiplying by $|a|$ both sides and letting $|a| \to \infty$ we get the said inequality for $C(F)$.

Hitting probability for open sets.

Let U be a bounded open set and define σ_U and $p_U(a)$ as in the case of compact sets F. Then $p_U(a)$ is harmonic outside ∂U and super harmonic in the whole space. Therefore $p_U(a) = \int_{\partial U} \frac{\mu_U(db)}{|a-b|}$ in $(\partial U)^c$, and $\mu_U(\partial U) = \lim_{|a| \to \infty} |a| \, p_U(a)$. Let $F_n \uparrow U$ be compact subsets of U. Then $C(F_n) = \lim_{|a| \to \infty} |\varepsilon| \, p_{F_n}(a)$ and the convergence is uniform in n since F_n are contained in a bounded set. Also since

$$p_U(a) = P_a(\sigma_U < \infty) = \lim_{n \to \infty} P_a(\sigma_{F_n} < \infty) = \lim p_{F_n}(a)$$

we have

$$\mu_U(\partial U) = \lim_{n \to \infty} C(F_n).$$

Therefore $\mu_U(\partial U)$ is the supremum of capacities of compact sets contained in U ; it is the <u>capacity</u> $C(U)$ of U by definition. Again $p_U(a) \leq \dfrac{C(U)}{\min_{b \in \partial U} |b-a|}$.

<u>Remark</u>. The capacity of any set is defined as follows. We have already defined the notion of capacity for compact sets. The capacity of any open set is by definition the supremum of the capacities of compact sets contained in it. The <u>outer capacity</u> of a set is the infimum of the capacities of open sets containing it, while the <u>inner capacity</u> is the supremum of the capacities of compact sets contained in it. If both are equal the set is called <u>capacitable</u> and the outer (or inner) capacity is called the <u>capacity</u> of the set. Choquet has proved that every Borel (even analytic) set is capacitable.

<u>Relative hitting probability</u> ($k \geq 1$).

Let F be a compact set contained in a bounded open set U and put $\sigma_{F/U} = \inf\{t : \tau_U > t > 0 \text{ and } x_t \in F\}$ where τ_U is the first leaving time from U. Let $p_{F/U}(a) = P_a(\sigma_{F/U} < \infty) = P_a\{\text{for some } t > 0, x_t \text{ reaches } F \text{ before it leaves } U\}$. $p_{F/U}(a)$ is called the (<u>relative</u>) <u>hitting probability</u> for F with respect to a and relative to U. Using the same idea as before we can prove

<u>Theorem 1'</u>

$p_{F/U}(a)$ is expressible as a potential induced by a bounded measure $\mu_{F/U}$ with the Green function $G_U(a,b)$, i.e.

$$p_{F/U}(a) = \int G_U(a,b)\, \mu_{F/U}(db) \ , \ a \in U,$$

where $\mu_{F/U}$ is concentrated on ∂F. Further $\mu_{F/U}$ is uniquely determined by F.

We can define the <u>relative capacity</u> $C_U(F)$ of F as $\mu_{F/U}(\partial F)$ and carry out similar discussions.

<u>Remark on absolute hitting probability</u> ($k \leq 2$).

In case $k = 1$, $p_F(a) \equiv 0$ or $\equiv 1$ according as $F \neq \emptyset$ or $= \emptyset$.

In case $k = 2$, we contend that $p_F(a) \equiv 1$ or 0 according as $C_U(F) > 0$ or $= 0$, where U is a bounded open set containing F. To prove this let V be another bounded open set such that $F \subset V \subset \bar{V} \subset U$. Let $\sigma_1(w) = \tau_U(w) + \sigma_V(w^+_{\tau_U})$, $\sigma_2(w) = \sigma_1(w) + \sigma_1(w^+_{\sigma_1})$, $\sigma_3(w) = \sigma_2(w) + \sigma_1(w^+_{\sigma_2}), \ldots$, $\sigma_n(w) = \sigma_{n-1}(w) + \sigma_1(w^+_{\sigma_{n-1}})$, etc. Then evidently $x_{\sigma_n} \in \partial V$ and $\sigma_n \uparrow \infty$; for, let $\sigma'_n(w) = \sigma_{n-1}(w) + \tau_U(w^+_{\sigma_{n-1}})$. Then $\sigma_{n-1} \leq \sigma'_n \leq \sigma_n$, and $x_{\sigma_n} \in \partial V$, $x_{\sigma'_n} \in \partial U$ so that if $\sigma_n \uparrow \sigma$, $x_\sigma \in \partial V \cap \partial U = \emptyset$ which is a contradiction. Hence $\sigma_n \uparrow \infty$ with P_a-probability 1. If $C_U(F) = 0$, then $p_{F/U}(x_{\sigma_n}) = 0$. Now

$$P_a(x_t \in F, \sigma_n < t \leq \sigma_{n+1}) = P_a(\sigma_F(w^+_{\sigma_n}) < \tau_U(w^+_{\sigma_n}))$$

$$= E_a(P_{x_{\sigma_n}}(\sigma_F(w) < \tau_U(w))) = E_a(p_{F/U}(x_{\sigma_n})) = 0.$$

Hence $p_F(a) = P_a(x_t \in F \text{ for some } t > 0) \leq \sum P_a(x_t \in F, \sigma_n < t \leq \sigma_{n+1}) = 0$.

Now

$$1-p_F(a) \leq P_a(x_t \notin F,\ 0<t<\sigma_n)$$

$$\leq P_a(\sigma_F(w^+_{\sigma_r}) > \tau_U(w^+_{\sigma_r})),\ 1 \leq r \leq n)$$

The set $(\sigma_F(w^+_{\sigma_r}) > \tau_U(w^+_{\sigma_r}),\ 1 \leq r \leq n-1)$ is $\mathcal{B}_{\sigma_{n^+}}$-measurable. For,

$$(\sigma_F(w^+_{\sigma_r}) < \tau_U(w^+_{\sigma_r})) = (\sigma_F\left[(w^-_{\sigma_{r+1}})^+_{\sigma_r(w^-_{\sigma_{r+1}})}\right] > \tau_U\left[(w^-_{\sigma_{r+1}})^+_{\sigma_r(w^-_{\sigma_{r+1}})}\right])$$

Hence $(\sigma_F(w^+_{\sigma_r}) < \tau_U(w^+_{\sigma_r})) \in \mathcal{B}_{\sigma_{r+1}} \subset \mathcal{B}_{\sigma_n}$, for $r+1 \leq n$.

[Note that if σ_1, σ_2 are two Markov times and $\sigma_1 < \sigma_2$, then $\mathcal{B}_{\sigma_1} \subset \mathcal{B}_{\sigma_2}$]. Hence by strong Markov property

$$P_a(\sigma_F(w^+_{\sigma_r}) > \tau_U(w^+_{\sigma_r}), 1 \leq r \leq n) = E_a\left[P_{x_{\sigma_n}}(\sigma_F > \tau_U) : \sigma_F(w^+_{\sigma_r}) > \tau_U(w^+_{\sigma_r}),\ 1 \leq r \leq n-1\right]$$

If $C_U(F) > 0$, since $p_{F/U}(a)$ is continuous on ∂V and always > 0 it has a minimum $\epsilon > 0$. Then

$$P_a(\sigma_F(w^+_{\sigma_r}) > \tau_U(w^+_{\sigma_r}), 1 \leq r \leq n) \leq (1-\epsilon) P_a(\sigma_F(w^+_{\sigma_r}) > \tau_U(w^+_{\sigma_r}), 1 \leq r \leq n-1)$$

$$\leq \ldots \leq (1-\epsilon)^n \longrightarrow 0.$$

This proves our contention.

§7. Regular points ($k \geq 3$).

In order to decide whether the generalised solution (the stochastic solution) $u(a) = u(a:f,U)$ satisfies the boundary conditions

$$\lim_{a \in U, a \to \xi} u(a) = f(\xi) , \quad \xi \in \partial U$$

we introduce the notion of regularity of boundary points.

Let U be an open set and $\xi \in \partial U$. Let

$$\tau_U^* = \inf \{ t : t > 0 \text{ and } x_t \notin U \},$$

and consider the event $\tau_U^* = 0$. This clearly belongs to $\mathcal{B}_{0+}$ and Blumenthal 0-1 law gives $P_\xi(\tau_U^* = 0) = 1$ or 0. If it is 1, ξ is called a <u>regular point</u> for U; if it is zero it is called <u>irregular</u> for U. Regularity is a local property. In fact, if ξ is regular for U, ξ is regular for $\Gamma \cap U$ for any open neighbourhood Γ of ξ and vice versa. We state here two important criteria for regularity.

<u>Theorem 1</u>. Let $\xi \in \partial U$.

(a) ξ is regular for U if and only if

$\lim_{a \in U, a \to \xi} P_a(x(\tau_U^*) \in \partial U \cap \Gamma) = 1$ for every open neighbourhood Γ of ξ

(2) ξ is irregular for U if and only if

$$\lim_{\Gamma \downarrow \xi} \underline{\lim}_{a \to \xi} P_a(x(\tau_U^*) \in \partial U \cap \Gamma) = 0.$$

<u>Theorem 2</u> (Wiener's test)

If $\xi \in \partial U$ and

$$F_n = (b : 2^{-(n+1)(k-2)} \leq |b - \xi| \leq 2^{-n(k-2)}, \ b \in U^c),$$

ξ is regular or irregular according as $\sum_n 2^{-n(k-2)} C(F_n) = \infty$ or $< \infty$.

We can prove the above two theorems using the same idea we used for the proof of Poincaré's test.

The following theorem, an immediate corollary of Theorem 1 gives the boundary values of the stochastic solution.

Theorem 3.

If U is a bounded open set, if ξ is regular for U and if f is bounded Borel on ∂U and continuous at ξ, then

$$\lim_{a \in U, a \to \xi} u(a:f,U) = f(\xi).$$

On the other hand if ξ is irregular for U, then there exist a continuous function f on ∂U such that the above equality is not true.

The following theorem, which we shall state without proof, shows that the set of irregular points is very small compared with the rest.

Theorem 4.

Let U be a bounded open set. Then the set of irregular points has capacity zero.

Using Theorems 3 and 4 we prove the following

Theorem 5.

If U is a bounded open set and if f is continuous on ∂U the stochastic solution $u(a) = u(a:f,U)$ is the unique bounded harmonic function defined in U such that

$$\lim_{a \in U, a \to \xi} u(a) = f(\xi), \quad \xi \in \partial U$$

except for a ξ-set of capacity zero.

Proof. It follows at once from Theorem 3 and 4 that the stochastic solution is a bounded harmonic function with boundary values f at regular points. Conversely let v be any bounded harmonic function with the boundary values f upto capacity zero. Let N be the set of all points ξ such that $v(a) \not\longrightarrow f(\xi)$. Then $C(N) = 0$ by assumption. Therefore there exists a decreasing sequence of open sets $G_m \supset N$ such that $\bar{G}_{m+1} \subset G_m$ and $C(G_m) \longrightarrow 0$. Since N is bounded we can assume that G_m are also bounded and since $N \subset \partial U$, we can assume that $\bigcap_m G_m \subset \partial U$. Let $a \in U$. Then $\rho(a, G_m) = \inf_{b \in G_m} \rho(a,b) >$ some positive constant for large m. Therefore

$$P_a(x_{\tau_U} \in N) \leq P_a(x_{\tau_U} \in \bigcap_m G_m) \leq P_a(\sigma_{G_m} < \infty) \leq \frac{C(G_m)}{(\rho(a,G_m))^{k-2}} \longrightarrow 0$$

so that $P_a(x_{\tau_U} \in N) = 0$. Let now U_n be open sets, $U_n \uparrow U$ such that $\bar{U}_n \subset U$ and every boundary point of U_n is a Poincaré point for U_n. Then $v(a) = E_a(b(x_{\tau_{U_n}}))$, $a \in U_n$ so that

$$v(a) = \lim_{n \to \infty} E_a(v(x_{\tau_{U_n}})) = E_a(\lim_{n \to \infty} v(x_{\tau_{U_n}}))$$

$$= E_a(\lim_{n \to \infty} v(x_{\tau_{U_n}}) : \lim_{n \to \infty} x_{\tau_{U_n}} = x_{\tau_U} \notin N)$$

$$= E_a(f(x_{\tau_U}) : x_{\tau_U} \notin N) = E_a(f(x_{\tau_U})) = u(a:f,U).$$

§8. Plane measure of a two-dimensional Brownian motion curve.

We have seen in Theorem 3 of §4 that the two-dimensional Brownian motion curve is dense in the plane. We now prove the following interesting theorem due to Paul Lévy.

Theorem 1 (P. Lévy).

The two dimensional Lebesgue measure of a two-dimensional Brownian motion curve is zero with probability 1 i.e. if $C(w) = \left\{x_s : 0 \leq s < \infty\right\}$, and $|C|$ = the Lebesgue measure of $C(w)$ then $P_a(|C| = 0) = 1$.

We first prove the following lemma.

Lemma. Let S be a Hausdorff space with the second countability axiom and W a class of continuous functions of $[0,t]$ into S. Let $\mathcal{B}$ be the Borel algebra generated by the class of all sets of the form $\left\{w : w \in W \text{ and } w(s) \in E\right\}$ where $0 \leq s \leq t$ and $E \in \mathcal{B}(S)$, $\mathcal{B}(S)$ being the class of Borel subsets of S (i.e. the Borel algebra generated by open sets of S). Let $C(w) = \left\{w(s) : 0 \leq s \leq t\right\}$. Then the function defined by

$$\begin{aligned} f(a,w) &= 1 \text{ if } a \in C(w) \\ &= 0 \text{ if } a \notin C(w) \end{aligned}$$

is $\mathcal{B}(S) \times \mathcal{B}$ -measurable in the pair (a,w).

Proof. It is clearly enough to prove that

$$\left\{(a,w) : a \notin C(w)\right\} \in \mathcal{B}(S) \times \mathcal{B}.$$

For any open set $U \subset S$ we have

$$(w : C(w) \subset U^c) = \bigcap_{\substack{r \leq t \\ r, \text{rational}}} \left\{ w : w(r) \in U^c \right\}$$

so that

$$(w : C(w) \subset U^c) \in \mathcal{B}.$$

Let now U_n be a countable base for S. Then it is not difficult to see that

$$\left\{(a,w) : a \notin C(w)\right\} = \bigcup_{n=1}^{\infty} \left[U_n \times \left\{w : C(w) \subset U_n^c \right\}\right]$$

using the fact that $C(w)$ being the continuous image of $[o,t]$ is closed. Q.E.D.

Proof of the theorem. To prove the theorem, it is enough to consider two-dimensional Brownian motion curves starting at zero i.e. a two-dimensional Wiener process. Let $x_t(w)$ be a two-dimensional Wiener process on $\Omega(\mathcal{B},P)$. It is enough to show that $E(|c_t|) = 0$, where $c_t = \{x_s : o \leq s \leq t\}$ and $|c_t|$ = the two-dimensional Lebesgue measure of c_t. From the lemma, the function $\chi(a,c_t)$ defined as

$$\chi(a,c_t) = 1 \text{ if } a \in c_t$$

$$= 0 \text{ if } a \notin c_t$$

is measurable in the pair (a,w). Since $|c_t| = \int_{R^2} \chi(a,c_t)da$, $|c_t|$ is measurable in w. Consider the following four processes:

1. $x_s(w)$, $o \leq s \leq t$
2. $y_s(w) = x_{s+t}(w) - x_t(w)$, $o \leq s \leq t$
3. $z_s(w) = x_{t-s}(w) - x_t(w)$, $o \leq s \leq t$
4. $u_s(w) = \dfrac{x_{2s}(w)}{\sqrt{2}}$, $o \leq s \leq t$.

All the four processes are continuous processes i.e. processes whose sample functions are continuous. Let

$$c_t^x = \{x_s : o \leq s \leq t\}$$

with similar meanings for c_t^u, c_t^y and c_t^z. Now the form of the Gaussian distribution shows that all the above four processes have the same joint distributions at any finite system of points. It follows that the distributions induced on $[R^2]^{[0,t]}$ by the above processes are the same. Also $\chi(a,c_t^x) = f(a,x)$ where f is the function in the lemma and x denotes the path. Thus we have

$$E(\chi(a,c_t^x)) = E(\chi(a,c_t^y)) = E(\chi(a,c_t^z)) = E(\chi(a,c_t^u)).$$

Hence

$$E(|c_t^x|) = \int_{R^2} E(\chi(a,c_t^x))da = \int_{R^2} E(\chi(a,c_t^u))da = E(|c_t^u|).$$

We have

$$c_{2t}^x = \{x_s : o \leq s \leq 2t\} = c_t^x \cup [c_t^y + x_t]$$

$$\equiv [c_t^x - x_t] \cup c_t^y = c_t^z \cup c_t^y ,$$

where $\equiv$ denotes congruency under translation. Therefore $|c_{2t}^x| + |c_t^y \cap c_t^z| = |c_t^z| + |c_t^y|$, and

$$E(|c_{2t}^x|) + E(|c_t^y \cap c_t^z|) = E(|c_t^z|) + E(|c_t^y|) = 2\,E(|c_t^x|).$$

Also

$$E(|c_{2t}^x|) = E(|\sqrt{2}\, c_t^u|) = E(2|c_t^u|) = 2\,E(|c_t^u|) = 2\,E(|c_t^x|)$$

Therefore

$$E(|c_t^y \cap c_t^z|) = 0 \text{ i.e. } \int_{R^2} E(\chi(a,c_t^z)\chi(a,c_t^y))da = 0.$$

Since the process y and z are easily seen to be independent

$$E(\chi(a,c_t^z)\chi(a,c_t^y)) = E(\chi(a,c_t^z))E(\chi(a,c_t^y)) = [E(\chi(a,c_t^x))]^2.$$

Therefore $\int \left[E(\chi(a,c_t^x))\right]^2 da = 0$ giving $E(\chi(a,c_t^x)) = 0$ for almost all a. Hence $\int E(\chi(a,c_t^x))da = 0$ i.e. $E(|c_t^x|) = 0$. This proves the theorem.

Section 4

Additive processes

§1. Definitions.

Let $x_. = (x_t, 0 \leq t < a)$ be a stochastic process on a probability space (Ω, P). If $I = (t_1, t_2]$, the increment of $x_.$ in I is by definition the random variable $x(I) = x_{t_2} - x_{t_1}$.

Definition. A process $x_. = (x_t)$ with $x_0 \equiv 0$ is called an additive (or differential) process, if for every finite disjoint system $I_1, \dots, I_n$ of intervals, $x(I_1), \dots, x(I_n)$ are independent.

We shall only consider additive processes $x_.$ for which $E(x_t^2) < \infty$ for all t. In this case $E(x_t) = m(t)$ exists and is called the first moment of x_t. $E((x_t - m(t))^2)$ is called the variance of x_t and is denoted by $V(x_t)$ or $v(t)$. If $y_t = x_t - m(t)$, $y_. = (y_t)$ is also additive.

Definition. A process $x_. = (x_t)$ is said to be continuous in probability at t_0 or said to have no fixed discontinuity at t_0, if for every $\epsilon > 0$,

$$\lim_{t \to t_0} P(|x_t - x_{t_0}| > \epsilon) = 0.$$

If it is continuous in probability at every point t it is said to be continuous in probability.

The following theorem is due to Doob.

Theorem 1.

If an additive process (x_t) has no fixed discontinuity then there exists a process (y_t) such that

1) $P(x_t = y_t) = 1$ for all t ;

2) almost all sample functions of (y_t) are d_1.

If further (y_t) , (y_t') are two such processes, then

$$P(\text{for every } t \text{ , } y_t = y_t') = 1.$$

$y_\cdot = (y_t)$ is called the standard modification of x_t. The proof can be seen in Doob's "Stochastic processes."

Definition. An additive process (x_t) with no point of fixed discontinuity and whose sample paths are d_1 with probability 1 is called a Lévy process.

It can be seen easily that Wiener processes and Poisson processes are particular cases of Lévy processes.

Definition. A process (x_t) is called temporally homogeneous if the probability distribution of $x_s - x_t$ $(s > t)$ depends only on $s-t$.

The above theorem of Doob shows that it is enough to study Lévy processes in order to study additive processes with no point of fixed discontinuity.

§2. Gaussian additive processes and Poisson additive processes.

The following two theorems give two elementary types of Lévy processes.

Definition. An additive process (x_t) with almost all sample paths continuous is called a Gaussian additive process. If for almost all w , the sample functions are step functions increasing with jump 1 the process is called a <u>Poisson additive process</u>.

We prove the following two theorems which justify the above nomenclature.

<u>Theorem 1</u>

Let (x_t) be a Lévy process. If $x_t(w)$ is continuous in t for almost all w, then $x(I)$ is a Gaussian variable.

The condition that x_t is continuous for almost all w is sometimes referred to as "(x_t) has no moving discontinuity" in contrast with "(x_t) has no fixed discontinuity".

<u>Proof</u>. Let $I = (t_0 , t_1]$. Since almost all sample functions are continuous, for any $\epsilon > 0$, there exists a $\delta(\epsilon) > 0$ such that

$$P(\text{for all } t,s \in I, |t-s| < \delta \Rightarrow |x_t - x_s| < \epsilon) > 1-\epsilon.$$

Noting this, let for each n,

$$t_0 = t_{n_0} < t_{n_1} < \ldots < t_{n_{p_n}} = t_1$$

be a subdivision of $(t_0, t_1]$, with $0 < t_{n_i} - t_{n_{i-1}} < \delta(\epsilon_n)$, where $\epsilon_n \downarrow 0$. Let $x_{nk} = x(t_{n_k}) - x(t_{n_{k-1}})$. Then $x = x(I) = \sum_{k=1}^{p_n} x_{nk}$. Define $x'_{nk} = x_{nk}$ if $|x_{nk}| < \epsilon_n$ and zero otherwise. Put $x_n = \sum_{k=1}^{p_n} x'_{nk}$. Then from the above it follows that

$$P(x = x_n) > 1-\epsilon_n ;$$

i.e. that $x_n \longrightarrow x$ in probability. Since x_{nk} are independent so are x'_{nk}. Therefore

$$E(e^{i\alpha x}) = \lim_{n\to\infty} E(e^{i\alpha x_n}) = \lim_{n\to\infty} \prod_{k=1}^{p_n} E(e^{i\alpha x'_{nk}}).$$

Let $m_{nk} = E(x'_{nk})$, $V_{nk} = V(x'_{nk})$, $m_n = \sum_{k=1}^{p_n} m_{nk}$ and $V_n = \sum_{k=1}^{p_n} V_{nk}$.

Then $|m_{nk}| \leq \epsilon_n$ and $V_{nk} \leq 4\epsilon_n^2$. Now

$$E(e^{i\alpha x}) = \lim_{n\to\infty} e^{i\alpha m_n} \prod_{k=1}^{p_n} E(e^{i\alpha(x'_{nk}-m_{nk})})$$

$$= \lim_{n\to\infty} e^{i\alpha m_n} \prod_{k=1}^{p_n} \left[1 - \frac{\alpha^2}{2} V_{nk}(1+O(\epsilon_n))\right],$$

so that

$$|E(e^{i\alpha x})| \leq \underline{\lim}_{n\to\infty} \prod_k e^{-\frac{\alpha^2}{2} V_{nk}} = \underline{\lim}_{n\to\infty} e^{-\frac{\alpha^2}{2} V_n} \leq e^{-\frac{\alpha^2}{2} \overline{\lim}\, V_n}.$$

Since $E(e^{i\alpha x})$ is continuous in α and is 1 at $\alpha = 0$, for sufficiently small α, $E(e^{i\alpha x}) \neq 0$. Hence $\overline{\lim}_{n\to\infty} V_n < \infty$, i.e. V_n is bounded. By taking a subsequence if necessary we can assume that $V_n \to V$.

We can very easily prove that if $z_n = \sum_{i=1}^{p_n} z_{ni}$ such that

1) $\sup_{1\leq i\leq p_n} |z_{ni}| \to 0$ as $n \to \infty$;

2) $\sum_{i=1}^{p_n} |z_{ni}|$ is bounded uniformly in n; and

3) $z_n \to z$, then

$$\lim_{n\to\infty} \prod_{i=1}^{p_n} [1-z_{ni}] = e^{-z}.$$

Now in our case $\max_k |V_{nk}| \leq 4\epsilon_n^2 \longrightarrow 0$, $\sum_{k=1}^{p_n} V_{nk}[1+O(\epsilon_n)] \longrightarrow V$ and $V_{nk} \geqslant 0$ so that

$$\lim_{n\to\infty} \prod_{k=1}^{p_n} \left[1-\frac{\alpha^2}{2} V_{nk}(1+O(\epsilon_n))\right] = e^{-\frac{\alpha^2}{2}V} .$$

Therefore $E(e^{i\alpha x}) = \lim_{n\to\infty} e^{i\alpha m_n} e^{-\alpha^2/V}$. This implies that $\varphi(\alpha) = \lim_{n\to\infty} e^{i\alpha m_n}$ exists. Now if $0 \leq \beta \leq \pi/2$,

$$\int_0^\beta \varphi(\alpha)d\alpha = \lim_n \int_0^\beta e^{i\alpha m_n} d\alpha = \lim_{n\to\infty} \frac{e^{i\beta m_n}-1}{im_n} = 0$$

if $m_n \longrightarrow \pm\infty$, and then $\varphi(\alpha) = 0$ for almost all $\alpha \leq \pi/2$, i.e. $E(e^{i\alpha x}) = 0$ for almost all $\alpha \leq \pi/2$ and this is a contradiction. Therefore $m_n \longrightarrow m$ and

$$E(e^{i\alpha x}) = e^{i\alpha m - \alpha^2/2\, V} .$$

<u>Theorem 2</u>

Let (x_t) be a Lévy process. If almost all sample functions are step functions with jump 1, then $x(I)$ is a Poisson variable.

<u>Proof</u>. From the continuity in probability of x_t ,

$$\sup_{|t-s|<n^{-1}, t_o\leq t,s\leq t_1} P(|x_t-x_s| \geqslant 1) \longrightarrow 0 \text{ as } n \longrightarrow \infty .$$

For each n , let $t_o = t_{no} < t_{n1} < \ldots < t_{np_n} = t_1$, $t_{ni} - t_{ni-1} \leq \frac{1}{n}$, be a subdivision of $[t_o,t_1]$ and let $x_{nk} = x_{tnk} - x_{tnk-1}$, $x'_{nk} = x_{nk}$ if $x_{nk} = 0$ or 1 and $x'_{nk} = 1$ if $x_{nk} \geqslant 2$. Put $x_n = \sum x'_{nk}$. Then since $P(x_n \longrightarrow x) = 1$,

$$E(e^{-\alpha x}) = \lim_{n\to\infty} E(e^{-\alpha x_n}) = \lim_{n\to\infty} \prod_{k=1}^{p_n} E(e^{-\alpha x'_{nk}})$$

$$= \lim_{n\to\infty} \prod_{k=1}^{p_n} \left[(1-p_{nk})+p_{nk}e^{-\alpha}\right] = \lim_{n\to\infty} \prod_{k=1}^{p_n} \left[1-p_{nk}(1-e^{-\alpha})\right]$$

$$\leq \underline{\lim}_{n\to\infty} \prod_{k=1}^{p_n} e^{-p_{nk}(1-e^{-\alpha})} = \underline{\lim}_{n\to\infty} e^{-P_n(1-e^{-\alpha})} = e^{-(1-e^{-\alpha})\overline{\lim}\, P_n},$$

where $p_{nk} = P(x_{nk} \geqslant 1) = P(x'_{nk} = 1)$ and $P_n = \sum_{k=1}^{p_n} p_{nk}$. Therefore P_n is bounded. We can assume that $P_n \longrightarrow p$. Again, since $\max_{1 \leq k \leq p_n} p_{nk} \longrightarrow 0$, $E(e^{-\alpha x}) = e^{-p(1-e^{-\alpha})}$.

§3. Lévy's canonical form.

Before considering the decomposition of a Lévy process, we prove some lemmas.

Lemma 1. Let (x_t) be a Lévy process and (y_t) a Poisson additive process. Suppose further that $(z_t) = ((x_t, y_t))$ is a vector-valued additive process. Then, if

$$P(\text{for every } t, x_t = x_{t-} \text{ or } y_t = y_{t-}) = 1,$$

the processes (x_t) and (y_t) are independent.

Proof. It is enough to prove that

$$P(x(I) \in E, y(I) \in F) = P(x(I) \in E) P(y(I) \in F).$$

For once this is proved we have, by the additivity of (z_t), for any finite disjoint system $I_1,\dots,I_n$ of intervals,

$$P(x(I_i)\in E_i, y(I_i)\in F_i\ ,\ i=1,2,\dots,n) = \prod_{i=1}^{n} P(x(I_i)\in E_i, y(I_i)\in F_i)$$

$$= \prod_{i=1}^{n} P(x(I_i)\in E_i)P(y(I_i)\in F_i)$$

$$= P\left[x(I_i)\in E_i, i=1,2,\dots,n\right] P\left[y(I_i)\in F_i, i=1,2,\dots,n\right],$$

and the proof can be completed easily.

Since $y(I)$ is a Poisson variable, it is enough to prove that $E(e^{i\alpha x(I)}: y(I) = K) = E(e^{i\alpha x(I)})P(y(I) = K)$.

Let $I = (t_o, t_1]$. For each n let $t_o = t_{no} < t_{n_1} < \dots < t_{nn} = t_1$, $t_{ni} - t_{n,i-1} = \frac{1}{n}(t_1 - t_o)$ be the subdivision of I into n equal intervals. Put $x_{ni} = x(t_{ni}) - x(t_{ni-1})$, $y_{ni} = y(t_{ni}) - y(t_{ni-1})$ $x'_{ni} = x_{ni}$ if $y_{ni} = 0$, $x'_{ni} = 0$ if $y_{ni} \geqslant 1$, and $x_n = \sum_{i=1}^{n} x'_{ni} = \sum_{y_{ni}=0} x_{ni}$. We have $x = x(I) = \sum_{i=1}^{n} x_{ni}$ and $|x(w) - x_n(w)| \leqslant \sum_{y_{ni}(w)\geqslant 1} x_{ni}(w)$. Since $y_t(w)$ is a Poisson variable increasing with jump 1, the number of terms in the right hand side of the last inequality is at most $y(w) = p$ (say). Suppose that $\tau_1(w),\dots,\tau_p(w)$ are the points in I, at which $y_t(w)$ has jumps. Then $|x(w) - x_n(w)| \leq \sum_{j=1}^{p} |x(s'_{nj}) - x(s_{nj})|$ where $(s_{nj}, s'_{nj}]$ is the interval of the nth subdivision which contains $\tau_j(w)$. Now $|x(s'_{nj}) - x(s_{nj})| \leq |x(s'_{nj}) - x(\tau_j)| + |x(\tau_j) - x(s_{nj})|$. Since at $\tau_1,\dots,\tau_p$, $y_t(w)$ has jumps, $x_t(w)$ has no jumps at these points. Therefore $|x(\tau_j) - x(s_{nj})|$ and

$\| x(s'_{nj})-x(\tau_j)\| \to 0$ as $n \to \infty$. Thus $P(x_n \to x) = 1$. Now

$$E(e^{i\alpha x_n} : y = k)$$

$$= \sum_{r\le k} \sum_{\substack{0\le\lambda_1\le\lambda_2\le\ldots\le\lambda_r\le n \\ p_1+\ldots+p_r = k \\ p_1,\ldots,p_r \geq 1}} E\left(e^{i\alpha \sum_{\lambda\ne\lambda_\sigma} x_{n\lambda}} : \begin{array}{l} y_{n\lambda_\sigma} = p_\sigma, \sigma=1,2,..r \\ y_{n\lambda} = 0, \ \lambda\ne\lambda_\sigma \end{array}\right)$$

Put
$$E\left(e^{i\alpha \sum_{\lambda\ne\lambda_\sigma} x_{n\lambda}} : \begin{array}{l} y_{n\lambda_\sigma} = p_\sigma, \ 1\le\sigma\le r \\ y_{n\lambda} = 0 ; \ \lambda\ne\lambda_\sigma \end{array}\right) = E_{r(\lambda)(p)}$$

Using the hypothesis that (x_t,y_t) is addtive one shows without difficulty that

$$E_{r(\lambda)(p)} = \prod_{\lambda\ne\lambda_\sigma} E(e^{i\alpha x_{n\lambda}} : y_{n\lambda} = 0) \prod_{1\le\sigma\le r} P(y_{n\lambda_\sigma} = p_\sigma),$$

so that

$$E_{r(\lambda)(p)} = E(e^{i\alpha \sum_{\lambda\ne\lambda_\sigma} x_{n\lambda}} : y_{n\lambda} = 0, \lambda\ne\lambda_\sigma) P(y_{n\lambda_\sigma} = p_\sigma, 1\le\sigma\le r).$$

Also $P(y=0)=P(y_{n\lambda}=0 \text{ for all } \lambda)=P(y_{n\lambda}=0, \lambda\ne\lambda_\sigma)P(y_{n\lambda_\sigma}=0, 1\le\sigma\le r)$.
Therefore (using the additivity of (x_t,y_t) again)

$$E_{r(\lambda)(p)}P(y=0)=E(e^{i\alpha \sum_{\lambda\ne\lambda_\sigma} x_{n\lambda}} : y_{n\lambda}=0, \lambda\ne\lambda_\sigma) P(y_{n\lambda_\sigma}=0, 1\le\sigma\le r) \times$$

$$\times P(y_{n\lambda_\sigma}= p_\sigma, 1\le\sigma\le r)P(y_{n\lambda}=0, \ \lambda\ne\lambda_\sigma)$$

$$= E(e^{i\alpha \sum_{\lambda\ne\lambda_\sigma} x_{n\lambda}} : y_{n\lambda}=0 \text{ for all } \lambda)P(y_{n\lambda_\sigma}=p_\sigma, 1\le\sigma\le r, y_{n\lambda}=0, \ \lambda\ne\lambda_\sigma)$$

$$= E(e^{i\alpha \sum_{\lambda\ne\lambda_\sigma} x_{n\lambda}} : y=0)P(y_{n\lambda_\sigma}=p_\sigma, 1\le\sigma\le r, y_{n\lambda}=0, \ \lambda\ne\lambda_\sigma).$$

Therefore

$$P(y=0)E(e^{i\alpha x_n} : y=k) = \sum_{r\le k} \sum_{\substack{0\le\lambda_1\le\ldots<\lambda_r\le n \\ p_1+\ldots+p_r=k \\ p_1,p_2,\ldots,p_r\geqslant 1}} E_{(\lambda)(p)}\, P(y=0)$$

$$= \sum_{r\le k} \sum_{\substack{0\le\lambda_1<\ldots<\lambda_r\le n \\ p_1+\ldots+p_r=k \\ p_1,\ldots,p_r\geqslant 1}} E(e^{i\alpha\sum_{\lambda\ne\lambda_\sigma} x_{n\lambda}} : y=0)P(y_{n\lambda_\sigma}=p_\sigma, y_{n\lambda}=0, \lambda\ne\lambda_\sigma)$$

Now $\left|E(e^{i\alpha\sum_{\lambda\ne\lambda_\sigma} x_{n\lambda}} : y=0)-E(e^{i\alpha x} : y=0)\right| \le E(|e^{i\alpha\sum_{1\le\sigma\le r} x_{n\lambda_\sigma}} - 1|)$

$$\le \sum_{\sigma=1}^{r} E(|e^{i\alpha x_{n\lambda_\sigma}}-1|) \le K \sup_{\substack{|t-s|\le\frac{1}{n}(t_1-t_0) \\ t_0\le t,s\le t_1}} E(|e^{i\alpha x_t} - e^{i\alpha x_s}|) \longrightarrow 0,$$

as $n\longrightarrow\infty$, since x_t has no point of fixed discontinuity. We thus have, since $P(y=k) = \sum_{r\le k} \sum_{\substack{(\lambda)\\(p)}} P(y_{n\lambda_\sigma} = p_\sigma\,, y_{n\lambda} = 0, \lambda\ne\lambda_\sigma)$,

$$|P(y=0)E(e^{i\alpha x_n} : y=k)-P(y=k)E(e^{i\alpha x} ; y=0)|$$

$$\le \sum_{r\le k} \sum_{\substack{(\lambda)\\(p)}} \left|E(e^{i\alpha\sum_{\lambda\ne\lambda_\sigma} x_n} : y=0)-E(e^{i\alpha x} : y=0)\right| P(y_{n\lambda_\sigma}=p_\sigma, y_{n\lambda}=0, \lambda\ne\lambda_\sigma)$$

$$\le \sup_{(\lambda),(p)} \left|E(e^{i\alpha\sum_{\lambda\ne\lambda_\sigma} x_{n\lambda}} : y=0)-E(e^{i\alpha x} : y=0)\right| \sum_{r\le k} \sum_{\substack{(\lambda)\\(p)}} P(y_{n\lambda_\sigma} = p_\sigma, y_{n\lambda} = 0, \lambda\ne\lambda_\sigma)$$

$$\le \sup_{(\lambda),(p)} \left|E(e^{i\alpha\sum_{\lambda\ne\lambda_\sigma} x_{n\lambda}} : y=0) - E(e^{i\alpha x} : y=0)\right| \longrightarrow 0.$$

Therefore $p(y=0)E(e^{i\alpha x}: y=k) = E(e^{i\alpha x}: y = 0)P(y = k)$.

Summing the above for $k = 0,1,2,\ldots$ we get $P(y=0)E(e^{i\alpha x}) = E(e^{i\alpha x}: y=0)$. Hence finally we have

$$P(y=0)E(e^{i\alpha x}: y=k) = E(e^{i\alpha x}: y=0)P(y=k) = E(e^{i\alpha x})P(y=0)P(y=k),$$

i.e.
$$E(e^{i\alpha x}: y=k) = P(y=k)E(e^{i\alpha x}).$$

We have proved the lemma.

Remark. We can prove that if $x_{\cdot} = (x_t)$, $y_{\cdot} = (y_t)$ are independent Levy processes, then

$$P(\text{for every } t, x_t = x_{t-} \text{ or } y_t = y_{t-}) = 1.$$

Lemma 2 (Ottaviani). If $r_1(\cdot),\ldots,r_n(\cdot)$ are independent stochastic processes almost all of whose sample functions are of type d_1 , then for any $\epsilon > 0$,

$$P\left[\max_{1\leq m\leq n} \|r_1(\cdot)+\ldots+r_m(\cdot)\| > 2\epsilon\right] \leq \frac{P\left[\|r_1+\ldots+r_n\| > \epsilon\right]}{1-\max\limits_{1\leq m\leq n-1} P\left[\|r_{n+1}+\ldots+r_n\| > \epsilon\right]},$$

where $\|r\| = \|r(\cdot)\| = \sup\limits_{0\leq s\leq t} |r(s)|$.

Proof. Let

$$A_m = (\max_{1\leq \mu\leq m-1} \|r_1+\ldots+r_\mu\| \leq 2\epsilon, \|r_1+\ldots+r_m\| > 2\epsilon),$$

$$B_m = (\|r_{m+1}+\ldots+r_n\| \leq \epsilon).$$

Then since A_m are disjoint, $A_m \cap B_m$ are also disjoint. Further

$$\bigcup_{m=1}^{n} A_m \cap B_m \subset C = (\| r_1 + \ldots + r_n \| > \epsilon) \text{ , so that}$$

$$P(C) \geqslant \sum P(A_m \cap B_m) = \sum P(A_m) P(B_m) \geqslant P(\cup A_m) \min_{m=1}^{n} P(B_m)$$

If we now note that $\min_{m=1}^{n} P(B_m) = 1 - \max_{1 \leq m \leq n} P(B_m^c)$ we get the result.

Lemma 3. Let (x_t) be a Lévy process, such that $E(x(t)) = 0$, $E(x(t)^2) < \infty$. Then for any $\epsilon > 0$,

$$P\left[\sup_{0 \leq s \leq t} |x(s)| > \epsilon\right] < \frac{1}{\epsilon^2} E(x(t)^2).$$

Proof. This lemma is the continuous version of Kolmogoroff's inequality which is as follows.

Kolmogoroff's inequality. If $x_1, \ldots, x_n$ are independent random variables with $E(x_i) = 0$, $E(x_i^2) < \infty$, $i=1,2,\ldots,n$, and if $S_m = x_1 + \ldots + x_m$, then

$$P(\max_{1 \leq m \leq n} |S_m| > \epsilon) \leqslant \frac{1}{\epsilon^2} E(S_n^2).$$

The lemma follows easily from this inequality.

Let now $(x_t, 0 \leq t < a)$ be a Lévy process, $S = \{(s,u) : 0 \leq s < a, -\infty < u < \infty\}$. Let $\mathcal{B}(S)$ be the set of Borel subsets of S and

$$\mathcal{B}^+(S) = (E : E \in \mathcal{B}(S) \text{ and } \rho(E, \text{s-axis}) > 0).$$

For every w we define

$$J(w) = ((t,u) \in S : x_t(w) - x_{t-}(w) = u \neq 0 , 0 \leq t < a).$$

For $E \in \mathcal{B}(S)$ put $p(E)$ = number of points in $J(w) \cap E$. For fixed

w, therefore p is a measure on $\mathcal{B}(S)$. We can prove that $p(E)$ is measurable in w, for fixed $E \in \mathcal{B}^+(S)$. Let $\sigma(M) = E(p(M))$ for $M \in \mathcal{B}^+(S)$. Then we have the

<u>Theorem</u>.

$$x_t = x_\infty(t) + \lim_{n\to\infty} \int_{[o,t]\times(u:1\geq|u|>\frac{1}{n})} [u\,p(ds\,du) - u\,\sigma(ds\,du)] + \int_{[o,t]\times(|u|>1)} up(ds\,du)$$

where $x_\infty(t)$ is continuous.

<u>Proof</u>. The proof is in several stages. Let $E_t = E \cap [(s,u) : o \leq s \leq t]$ for $E \in \mathcal{B}^+(S)$.

1. We shall first prove that $y_t^E = p(E_t)$ is an additive Poisson process.

Using the fact that x_t is of type d_1 it is not difficult to see that $y_t^E < \infty$, and that it increases with jump 1.

Let $\mathcal{B}_{ts}$ be the least Borel algebra with respect to which $x_u - x_v$, $s \leq u,v \leq t$, are measurable. We shall prove that $y_t^E - y_s^E$ is $\mathcal{B}_{ts}$-measurable. It suffices to prove this when $E = G$ is open. Let $G_m \uparrow G$, $\bar{G}_m \subset G_{m+1}$ be a sequence of open sets such that $\bar{G}_m$ is compact. Let $y_t^G - y_s^G = N$, $y_t^{G_m} - y_s^{G_m} = N_m$. For every n let $t_k^n = s + k\frac{(t-s)}{n}$, $k = 1,2,\ldots,n$ and N_n^m = number of k such that $(t_k^n, x_{t_k^n} - x_{t_{k-1}^n}) \in G_m$. Then $N_{m-1} \leq \overline{\lim}_{n\to\infty} N_n^m \leq N_{m+1}$, N_n^m is measurable in w with respect to $\mathcal{B}_{ts}$, and $y_t^G - y_s^G = \lim_{m\to\infty} \overline{\lim}_{n\to\infty} N_n^m$.

Now suppose that $I_i = (s_i, t_i]$ $i=1,2,\ldots,n$ are disjoint. Then $\mathcal{B}_{t_i s_i}$, $1 \leq i \leq n$ are independent and $y_{I_i}^E$ is $\mathcal{B}_{t_i s_i}$-measurable.

Therefore y_t^E is an additive process.

Finally y_t^E has no fixed discontinuity. For, a fixed discontinuity of y_t^E is also a fixed discontinuity of x_t.

Thus we have proved that $p(E_t)$ is an additive Poisson process.

2. Let $$r(E_t) = \sum_{(s,u)\in E_t\cap J} u = \int_{E_t} u\, p(ds\, du).$$

We prove that $r(E_t)$ is additive. For every $w \in \Omega$, p is a measure on $\mathcal{B}(E_t - E_s)$. Any simple function on $E_t - E_s$ is of the form $\sum_{i=1}^n a_i \chi_{F_i}$ where $F_i \in \mathcal{B}(E_t - E_s), i=1,2,\ldots,n$, are disjoint. Also $\int_{E_t - E_s} (\sum a_i \chi_{F_i}) p(ds\, du) = \sum a_i\, p(F_i)$, so that $\int_{E_t-E_s} (\sum a_i \chi_{F_i}) p(ds du)$ is $\mathbb{B}_{ts}$-measurable. It follows that

$$r(E_t) - r(E_s) = \int_{E_t - E_s} u p(ds\, du)$$

is $\mathcal{B}_{ts}$-measurable. Let $x_t^E = x_t - r(E_t)$. Using the fact that $r(E_t) - r(E_s)$ is $\mathcal{B}_{ts}$-measurable, it is seen without difficulty that x_t^E is a Lévy process. Since $z_t^E = (x_t^E, y_t^E)$ is additive, and $P[x_t^E = x_{t-}^E$ or $y_t^E = y_{t-}^E$ for every $t] = 1$ it follows that $x_\cdot^E$ and $y_\cdot^E$ are independent.

3. Now we prove that if $E_1, \ldots, E_n \in \mathcal{B}^+(S)$ are disjoint then $x_\cdot^{E_1 \cup \cdots \cup E_n}, y_\cdot^{E_1}, \ldots, y_\cdot^{E_n}$ are independent. For simplicity we prove this for $n=2$. Put $x_t' = x_t^{E_1}$. Then $(x_t')^{E_2} = x_t^{E_1 \cup E_2}$ and the process $y_\cdot^{E_2}$ defined with respect to x_t' is the same as $y_t^{E_2}$ with respect to x_t. Hence, since $(x_\cdot')^{E_2}$ and $y_\cdot^{E_2}$ are independent from

2, $x_{.}^{E_1 \cup E_2}$ and $y_{.}^{E_2}$ are independent. Further $(x_{.}^{E_1 \cup E_2}, y_{.}^{E_2})$ is measurable with respect to $\mathcal{B}(x_{.}^{E_1})$, the least Borel algebra with respect to which $x_t^{E_1}$ is measurable for all t, and $\mathcal{B}(x_{.}^{E_1})$, $\mathcal{B}(y_{.}^{E_1})$ are independent. Therefore $(x_{.}^{E_1 \cup E_2}, y_{.}^{E_2})$ and $y_{.}^{E_1}$ are independent. It follows that $x_{.}^{E_1 \cup E_2}$, $y_{.}^{E_1}$ and $y_{.}^{E_2}$ are independent.

4. x_t^E and $r(E_t)$ are independent.

Since $r(E_t) = \int_{E_t} u\, p(ds\, du)$, it is enough to prove that if F is a simple function on E_t, $\int_{E_t} F\, p\,(ds\, du)$ and x_t^E are independent; this follows from 3.

5. If $\sigma(M) = E(p(M))$ then $E(e^{i\alpha r(E_t)}) = \exp(\int_{E_t} (e^{i\alpha u}-1)\sigma(dsdu))$.

It is again enough to prove this for simple functions on E_t. Note that if y is a Poisson variable then $E(e^{i\alpha y}) = e^{\lambda(e^{i\alpha}-1)}$ where $\lambda = E(y)$, so that for any β we have $E(e^{i\alpha\beta y}) = e^{\lambda(e^{i\alpha\beta}-1)}$.

Let $f = \sum s_i \chi_{F_i}$ be a simple function on E_t with $F_i, 1 \leq i \leq n$ disjoint. Since $p(F_i)$ are independent random variables we have

$$E(\exp(\int_{E_t} f\, p(dsdu))) = E(e^{i\alpha \sum_{j=1}^{n} s_j\, p(F_j)}) = \prod_{j=1}^{n} E(e^{i\alpha s_j\, p(F_j)})$$

$$= \prod_{1 \leq j \leq n} \exp(\sigma(F_j)(e^{i\alpha s_j}-1)) = \prod_{1 \leq j \leq n} \exp(\int_{E_t} (e^{i\alpha \chi_{F_j} s_j}-1)\sigma(dsdu))$$

$$= \exp(\int_{E_t} (e^{i\alpha \sum_{j=1}^{n} s_j \chi_{F_j}} -1)\, \sigma(ds\, du)) = \exp(\int_{E_t} (e^{i\alpha f}-1)\sigma(dsdu)).$$

6. Let $U = ((s,u) \in S : |u| > 1)$, $U^n = ((s,u) \in S : \frac{1}{n} \leq |u| \leq 1)$.

Then $x_t = x_t^{U^n} + r(U_t^n)$, and since (x_t^E) and $r(E_t)$ are independent

$$|E(e^{i\alpha x_t})| = |E(e^{i\alpha x_t^{U^n}})||E(e^{i\alpha r(U_t^n)})| \leq |E(e^{i\alpha r(U_t^n)})| =$$

$$= \left|\exp\left(\int_{U_t^n} (e^{i\alpha u} - 1)\sigma(ds\,du)\right)\right| = \exp\left(\int_{U_t^n} (\cos\alpha u - 1)\sigma(dsdu)\right)$$

$$\leq \exp\left(-\frac{\alpha^2}{4}\int_{U_t^n} u^2\,\sigma(ds\,du)\right),$$

because $\cos\alpha u - 1 \leq -\frac{\alpha^2 u^2}{4}$ for $|\alpha| \leq 1$. It follows that $\int_{U_t^n} u^2\sigma(ds\,du) < \infty$ for every n. Therefore $\lim_{n\to\infty}\int_{U_t^n} u^2\sigma(dsdu) < \infty$.

7. Let $r_n(t) = r(U_t^n) - E(r(U_t^n))$, then $r_n(t)$ converges uniformly in $[o,a)$. The limit we denote by $r_\infty(t)$.

Now $r(U_t^{m+k+1}) - r(U_t^{m+k}) = r(U_t^{m+k+1} - U_t^{m+k})$. It follows that $r_{m+k}(\cdot) - r_{m+k-1}(\cdot)$, $k = 1,2,\ldots,n-m$ are independent. Using Lemmas 2 and 3,

$$P\left(\max_{1\leq k\leq n-m} \|r_{m+k} - r_m\| > 2\epsilon\right) \leq \frac{P(\|r_n - r_m\| > \epsilon)}{1 - \max\limits_{1\leq k\leq n-m-1} P(\|r_n - r_{m+k}\| > \epsilon)}$$

$$\leq \frac{\frac{1}{\epsilon^2}\int_{U_t^n - U_t^m} u^2\sigma(ds\,du)}{1 - \frac{1}{\epsilon^2}\int_{U_t^n - U_t^m} u^2\sigma(ds\,du)} \longrightarrow 0 \text{ as } m,n \longrightarrow \infty.$$

since

$$E(|r_n(t)-r_m(t)|^2) = E(|r(U_t^n - U_t^m) - E(r(U_t^n - U_t^m)|^2)$$

$$= E\left[\left(\int_{U_t^n - U_t^m} u\left[p(ds\,du) - \sigma(ds\,du)\right]\right)^2\right]$$

and

$$E\left[\left(\int_{E_t} u(p(ds\,du) - \sigma(ds\,du)\right)^2\right] = \int_{E_t} u^2\sigma(ds\,du)$$

which can be proved by first considering simple functions etc., and noting the fact that if y is a Poisson variable, then

$$E\left[(y-E(y))^2\right] = E(y).$$

8. Let $x_n(t) = x_t^{U^n} + E(r(U_t^n)) - r(U_t) = x_t - r_n(t) - r(U_t)$. Since $r_n(t)$ converges uniformly, in every compact subinterval of $[o,a)$, with probability 1, $x_n(t)$ converges uniformly in $[o,a)$, say, to $x_\infty(t)$. Since $x_n(t)$ has no jumps exceeding $\frac{1}{n}$ in absolute value $x_\infty(t)$ is continuous. We have

$$x_t = r(U_t) + \lim_{n\to\infty} r_n(t) + \lim_{n\to\infty} x_n(t) = \int_{U_t} u\, p(dsdu) + x_\infty(t)$$

$$+ \lim_{n\to\infty} \int_{U_t^n} \left[u\, p(dsdu) - u\sigma(dsdu)\right],$$

since $E(r(u_t^n)) = \int_{U_t^n} u\,\sigma\,(ds\,du)$. The theorem is proved.

Since $\int_{U_t} \sigma(ds\,du) = E(p(U_t)) < \infty$, $\int_{U_t} \frac{u}{1+u^2}\sigma(dsdu) < \infty$.

We have seen that $\lim_{n\to\infty} \int_{U_t^n} u^2\sigma(ds\,du) < \infty$. Therefore $\lim_{n\to\infty} \int_{U_t^n} \frac{u^3}{1+u^2}\sigma(ds\,du) < \infty$ and we can also write the last equation as

$$x_t = g(t) + \lim_{n\to\infty} \int_{[o,t]\times(u:\,|u|>\frac{1}{n})} \left[u\, p(dsdu) - \frac{u}{1+u^2}\sigma(dsdu)\right]$$

where $\quad g(t) = x_\infty(t) + \int_{U_t} \frac{u}{1+u^2}\sigma(dsdu) - \lim_{n\to\infty} \int_{U_t^n} \frac{u^3}{1+u^2}\sigma(dsdu).$

For simplicity we shall write

$$x_t = g(t) + \int_{s=o}^{t} \int_{-\infty}^{\infty} \left[u\, p(dsdu) - \frac{u}{1+u^2}\; \sigma(dsdu)\right].$$

In the general case when $x_o \neq 0$, we have

$$x_t = x_o + g(t) + \int_{s=o}^{t} \int_{-\infty}^{\infty} \left[u\, p(ds\, du) - \frac{u}{1+u^2}\; \sigma(dsdu)\right].$$

From now on we shall write

$$x_t = \int_{-\infty}^{\infty} u\, p([o,t]\times du) - \frac{u}{1+u^2}\, \sigma([o,t]\times du) \; + \; g(t).$$

Since x_t has no fixed discontinuity $P(|x_t - x_{t-}| > 0) = 0$. It follows that $\sigma(\{t\}\times U) = 0$. Noting this, it is not difficult to see that $\int_{U_t} \frac{u}{1+u^2}\, \sigma(ds\, du)$ and $\lim_{n\to\infty} \int_{U_t^n} \frac{u^3}{1+u^2}\, \sigma(ds\, du)$ are both continuous in t. Therefore $g(t)$ is continuous hence is a Gaussian additive process. Further we can show that $g(t)$ and $\int_{-\infty}^{\infty} [u\, p([o,t]\times du) - \frac{u}{1+u^2}\, \sigma([o,t]\times du)$ are independent. We have

$$E(e^{i\alpha(x_t-x_s)}) = E(\exp(i\alpha\int_{-\infty}^{\infty}[u\ p([s,t]\times du) - \frac{u}{1-u^2}\sigma([s,t]\times du)]))\times$$

$$\times E(e^{i\alpha[g(t)-g(s)]})$$

$$= \lim_{n\to\infty} E(\exp(i\alpha\int_{|u|>\frac{1}{n}}[u\ p([s,t]\times du) - \frac{u}{1+u^2}\sigma([s,t]\times du)]))\times$$

$$\times \exp(i(m(t)-m(s))\alpha - \frac{v(t)-v(s)}{2}\alpha^2)$$

$$= \lim_{n\to\infty} \exp\Big[\int_{|u|>\frac{1}{n}}(e^{i\alpha u} - 1-\frac{i\alpha u}{1+u^2})\sigma([s,t]\times du)\Big]\times$$

$$\times \exp(i\alpha(m(t)-m(s)) - \frac{v(t)-v(s)}{2}\alpha^2).$$

Therefore

$$\log E(e^{i\alpha(x_t-x_s)}) = i\alpha[m(t)-m(s)] - \frac{v(t)-v(s)}{2}\alpha^2 +$$

$$+ \int_{-\infty}^{\infty}[e^{i\alpha u}-1-\frac{i\alpha u}{1+u^2}\ \sigma([s,t]\times du).$$

Since $g(t)$ is Gaussian $m(t)$ and $v(t)$ are continuous in t and $v(t)$ increases with t.

Conversely, given m, v and σ such that 1) $m(t)$ is continuous in t, 2) $v(t)$ is continuous and increasing, 3) $\sigma[\{t\}\times U] = 0$, $\int_{-\infty}^{\infty}\frac{u^2}{1+u^2}\sigma([0,t]\times du) < \infty$, we can construct a unique (in law) Lévy process.

Let us now consider some special cases. If $\int_{-\infty}^{\infty} u^2\sigma([o,t] \times du) < \infty$ we can write

$$x_t = g_1(t) + \int_{-\infty}^{\infty} u[p([o,t]\times du) - \sigma([o,t]\times du) .$$

The condition $\int_{-\infty}^{\infty} \frac{|u|}{1+|u|} \sigma([o,t] \times du) < \infty$ is equivalent to the two conditions 1) $\int_{-\infty}^{\infty} \frac{u^2}{1+u^2} \sigma([o,t]\times du) < \infty$ and 2) $\int_{-\infty}^{\infty} \frac{|u|}{1+u^2} \sigma([o,t]\times du) < \infty$, so that if $\int_{-\infty}^{\infty} \frac{|u|}{1+|u|} \sigma([o,t]\times du) < \infty$ we can write

$$x_t = \int_{-\infty}^{\infty} u\, p([o,t] \times du) + g_2(t)$$

and

$$\log E(e^{i\alpha x_t}) = i\alpha m(t) - \frac{v(t)}{2}\alpha^2 + \int_{-\infty}^{\infty} [e^{i\alpha u} - 1] \sigma([o,t] \times du).$$

The condition $\int_{-\infty}^{\infty} \frac{u^2}{1+|u|} \sigma([o,t] \times du) < \infty$ is equivalent to the two conditions 1) $\int_{-\infty}^{\infty} \frac{u^2}{1+u^2} \sigma([o,t]\times du) < \infty$ and 2) $\int_{-\infty}^{\infty} \frac{u^3}{1+u^2} \sigma([o,t] \times du) < \infty$. Therefore if $\int_{-\infty}^{\infty} \frac{u^2}{1+|u|} \sigma([o,t]\times du) < \infty$ we can write

$$\log E(e^{i\alpha x_t}) = i\alpha m(t) - \frac{v(t)}{2}\alpha^2 + \int_{-\infty}^{\infty} \left[e^{i\alpha u} - 1 - \frac{i\alpha u}{1+u^2}\right] \sigma([o,t]\times du).$$

<u>Lemma</u>. If $f(\alpha) = i\, m\alpha - \frac{v}{2}\alpha^2 + \int_{-\infty}^{\infty} \left[e^{i\alpha u} - 1 - \frac{i\alpha u}{1+u^2}\right] \sigma(du) \equiv 0$,

where m and v are real and σ is a signed measure such that $\int_{-\infty}^{\infty} \frac{u^2}{1+u^2} \sigma(du) < \infty$, then $m = v = \sigma = 0$.

<u>Proof</u>. We have $0 \equiv f(\alpha)-\frac{1}{2}\int_{\alpha-1}^{\alpha+1} f(\beta)d\beta = \frac{v}{3} + \int_{-\infty}^{\infty} e^{i\alpha u}\left[1-\frac{\sin u}{u}\right]\sigma(du)$

so that if δ_o is the Dirac measure at 0,

$$\int_{-\infty}^{\infty}\left[\frac{v}{3}\,\delta_o(du) + \left(1-\frac{\sin u}{u}\right)\sigma(du)\right]e^{i\alpha u} \equiv 0.$$

It follows that $\frac{v}{3}\,\delta_o(A) + \int_A \left(1-\frac{\sin u}{u}\right)\sigma(du) = 0$. Taking $A = \{0\}$, since $\int_{\{o\}} \left(1-\frac{\sin u}{u}\right)\sigma(du) = 0$ we see that $v = 0$. It then follows that $\sigma = 0$ and hence $m = 0$.

From this lemma we can easily deduce that in the expression
$\log E(e^{i\alpha x_t}) = i\alpha m(t) - \frac{v(t)}{2}\alpha^2 + \int_{-\infty}^{\infty}\left[e^{i\alpha u}-1-\frac{i\alpha u}{1+u^2}\right]\sigma([o,t]\times du)$,
$m(t)$, $v(t)$ and σ are unique.

§4. <u>Temporally homogeneous Lévy processes</u>.

We shall prove that if (x_t) is a temporally homogeneous Lévy process, then $\log E(e^{i\alpha x_t}) = t\psi(\alpha)$ where $\psi(\alpha) = im\alpha - \frac{v}{2}\alpha^2 + \int_{-\infty}^{\infty}\left[e^{i\alpha u}-1-\frac{i\alpha u}{1+u^2}\right]\sigma(du)$.

<u>Definition</u>. Two random variables x and y on a probability space $\Omega(P)$ are said to be <u>equivalent in law</u> and we write $x \underset{L}{\sim} y$ if they yield the same distribution.

A stochastic process $(x_t, o \leq t < a)$ on Ω can be regarded as a measurable function into $R^{[o,a)}$. Two stochastic processes (x_t), (y_t), $o \leq t < a$ are said to be equivalent in law if they induce the same probability distribution on $R^{[o,a)}$ and we write $x_. \underset{L}{\sim} y_.$. If (x_t) and (y_t) are additive processes such that $x_t - x_s \underset{L}{\sim} y_t - y_s$, then we can prove that $x_. \underset{L}{\sim} y_.$.

Let D' denote the set of all d_1-type functions on $[o,a)$ into R^1. Then $D' \subset R^{[o,a)}$ and let $\mathcal{B}(D')$ be the induced Borel algebra on D' by $\mathcal{B}(R^{[o,a)})$. If $(x_t, o \leq t < a)$ is a Lévy process then the map $w \longrightarrow x_{.}(w)$ into D' is measurable; also if $x_t^{(h)} = x_{t+h} - x_t$, $o \leq t < a-h$ we can show that (x_t) is temporally homogeneous if and only if $x_{.} \underset{L}{\sim} x_{.}^{(h)}$.

Now consider D'. Let $E \in \mathcal{B}^+(S)$,

$$J(f) = \left\{(s,u) : f(s) - f(s-) = u \neq 0\right\}, \; f \in D'$$

and $F_t^E(f)$ = number of points in $J(f) \cap E_t$.

We can show that $F_t^E(f) < \infty$ and that F_t^E is measurable on D. The proof of measurability of F_t^E follows exactly on the same lines as that of the measurability of y_t^E. We have clearly $p(E_t) = y_t^E = F_t^E(x_.)$.

Let $E \in \mathcal{B}([o,t])$ and $U \in \mathcal{B}(R^1)$ be such that $E_1 = E \times U \in \mathcal{B}^+(S)$. If h is such that $t + h < a$ we prove that $\sigma((E+h) \times U) = \sigma(E \times U)$. Let $E_2 = (E+h) \times U$. Then $\sigma(E_2) = E(y_{t+h}^{E_2} - y_h^{E_2})$. Let $x_t^{(h)} = x_{t+h} - x_h$. Since x_t is temporally homogeneous $x_{.}^{(h)} \underset{L}{\sim} x_.$. Also $y_{t+h}^{E_2} - y_h^{E_2} = F_t^{E_1}[x_{.}^{(h)}]$ and $y_t^{E_1} = F_t^{E_1}[x_.]$. It follows that $E(y_t^{E_1}) = E(y_{t+h}^{E_2})$. Thus for fixed U, σ is a translation-invariant measure on $\mathcal{B}[(o,a))]$ and hence is the Lebesgue measure, i.e. $\sigma(E \times U) = m(E)\,\sigma_1(U)$, where $m(E)$ is the Lebesgue measure of E and $\sigma_1(U)$ is a constant depending on U. Since σ is a measure on R^2, it follows that σ_1 is also a measure. Hence $\sigma(ds\,du) = ds\,\sigma_1(du)$. We shall drop the suffix 1 and use the same symbol σ. Thus

$$\int_0^t \int_{-\infty}^{\infty} \left[e^{i\alpha u} - 1 - \frac{i\alpha u}{1+u^2}\right] \sigma(ds du) = t \int_{-\infty}^{\infty} \left[e^{i\alpha u} - 1 - \frac{i\alpha u}{1+u^2}\right] \sigma(du).$$

Now $\log E(e^{i\alpha(x_t - x_s)})$ depends only on $t-s$. Therefore $m(t)-m(s)$ and $v(t)-v(s)$ depend only on $t-s$. Hence $m(t) = m\,t$, $v(t) = v\,t$. Therefore, finally,

$$\log E(e^{i\alpha x_t}) = im\alpha t - \frac{vt}{2}\alpha^2 + t \int_{-\infty}^{\infty} \left[e^{i\alpha u} - 1 - \frac{i\alpha u}{1+u^2}\right] \sigma(du).$$

We shall now consider some special cases of temporally homogeneous Lévy process. We have seen that

$$x_t = g(t) + \int_{-\infty}^{\infty} \left[u\, p_t(du) - \frac{u}{1+u^2}\, t\, \sigma(du)\right],$$

where $p_t(du) = p([o,t] \times du)$. Since $g(t)$ is Gaussian additive and temporally homogeneous $g(t) = m\,t + \sqrt{v}\, B_t$, where B_t is a Wiener process. Thus

$$x_t = m\,t + \sqrt{v}\, B_t + \int_{-\infty}^{\infty} \left[u\, p_t(du) - \frac{u}{1+u^2}\, t\, \sigma(du)\right]$$

and $$\psi(\alpha) = i\,m\alpha - \frac{v}{2}\alpha^2 + \int_{-\infty}^{\infty} \left[e^{i\alpha u} - 1 - \frac{i\alpha u}{1+u^2}\right] \sigma(du).$$

Special cases: 1. $\sigma \equiv 0$. Then $x_t = m\,t + \sqrt{v}\, B_t$.

2. In case $m' = \lim_{\epsilon \downarrow 0} \int_{|u| > \epsilon} \frac{-u}{1+u^2} \sigma(du)$ exists, we can write

$x_t = (m+m')t + \sqrt{v}\, B_t + \int_{-\infty}^{\infty} u\, p_t(du)$, and

$$\psi(\alpha) = i(m+m')\alpha - \frac{v}{2}\alpha^2 + \int_{-\infty}^{\infty} \left[e^{i\alpha u} - 1\right] \sigma(du).$$

Note that if σ is symmetric, $m' = 0$.

3. If $m+m' = 0$ and $v = 0$, $x_t = \lim_{\varepsilon \downarrow 0} \int_{|u|>\varepsilon} u\, p_t(du)$,

$$\psi(\alpha) = \lim_{\varepsilon \downarrow 0} \int_{|u|>\varepsilon} [e^{-i\alpha u}-1]\, \sigma(du).$$

Such a process is called a <u>pure jump process</u>.

4. If $\lambda = \sigma(R^1) < \infty$, then

$$x_t = \int_{-\infty}^{\infty} u\, p_t(du), \psi(\alpha) = \lambda \int_{-\infty}^{\infty} [e^{i\alpha u}-1]\, \Theta\,(du) = \lambda\,[\theta(\alpha)-1],$$

where $\Theta(E) = \lambda^{-1}\sigma(E)$ and $\theta(\alpha)$ is the characteristic function of Θ. We have

$$E(e^{i\alpha x_t}) = e^{t\psi(\alpha)} = e^{-\lambda t} \sum_k \frac{t^k \lambda^k}{k!}\, \theta(\alpha)^k$$

$$= e^{-\lambda t} \sum_k \frac{t^k \lambda^k}{k!} \cdot [\text{characteristic function of } \Theta^{*k}] ,$$

where Θ^{*k} denotes the k-fold convolution of Θ. Since $E(e^{i\alpha x_t})$ is the characteristic function of the measure $\varphi(t,.)$, we have

$$\varphi(t,E) = e^{-\lambda t} \sum_k \frac{\lambda^k t^k}{k!}\, \Theta^{*k}\ (E).$$

<u>Remark</u>. If $\varphi(t,E) = P(x_t \in E)$ is symmetric, i.e. if $P(x_t \in E) = P(-x_t \in E)$, then $E(e^{i\alpha x_t})$ is real. Hence $\psi(\alpha)$ is real. Further, since $x_t \underset{L}{\sim} -x_t$, we have $x_{\cdot} \underset{L}{\sim} -x_{\cdot}$. It follows that $\sigma(db) = \sigma(-db)$. Therefore

$$\psi(\alpha) = i\, m\alpha - \frac{v}{2}\alpha^2 + 2 \int_0^{\infty} [\cos\alpha u - 1]\, \sigma(du).$$

Since $\psi(\alpha)$ is real, $m = 0$, so that

$$\psi(\alpha) = -\frac{v}{2}\alpha^2 + 2\int_0^\infty [\cos\alpha u - 1]\,\sigma(du).$$

§5. Stable processes.

Let $(x_t, 0 \le t < \infty)$ be a temporally homogeneous Lévy process. If $x_t \underset{L}{\sim} c_t x_1$, where c_t is a constant depending on t, we say that (x_t) is a stable process. We shall now give a theorem which characterises stable processes completely. From Lévy's canonical form, we have $E(e^{i\alpha x_t}) = e^{t\psi(\alpha)}$ where

$$\psi(\alpha) = i\,m\alpha - \frac{v}{2}\alpha^2 + \int_{-\infty}^{\infty} \left[e^{i\alpha u} - 1 - \frac{i\alpha u}{1+u^2}\right] \sigma(du).$$

Theorem 1.

$$\psi(\alpha) = \begin{cases} i\,m\alpha, m \text{ real}, \\ -a_o|\alpha|^2, \\ (-a_o + i\frac{\alpha}{|\alpha|}a_1)\cdot|\alpha|^c, \end{cases}$$

where $a_o > o$, $o < c < 2$ and a_1 is real.

Proof. Suppose that $\psi(\alpha)$ is not of the form $i\,m\alpha$.

We prove that if $\psi(c\alpha) = \psi(d\alpha)$, then $c = d$. For, if $\varepsilon = \min(\frac{c}{d}, \frac{d}{c}) < 1$ and $\psi(\alpha) = \psi(\varepsilon\alpha)$ so that $\psi(\alpha) = \psi(\varepsilon^n\alpha) \to 0$. Hence $\psi(\alpha) \equiv 0$ and this is the omitted case.

Since $e^{\psi(c_t\alpha)} = E(e^{ic_t\alpha x_1}) = E(e^{i\alpha x_t}) = e^{t\psi(\alpha)}$, we have $\psi(c_t\alpha) = t\psi(\alpha)$. Therefore

$$\psi(c_{ts}\alpha) = ts\psi(\alpha) = t\psi(c_s\alpha) = \psi(c_t c_s\alpha).$$

It follows that $c_{ts} = c_t\, c_s$.

We prove next that c_t is continuous. Let $c_{t_n} \to d$, as $t_n \to t$. If $d = \infty$, we should have, since $\psi(c_{t_n}\alpha) = t_n\psi(\alpha)$,

$$\psi(\alpha) = t_n\psi(c_{t_n}^{-1}\alpha) \to 0.$$

Therefore $d \neq \infty$ and $\psi(c_t\alpha) = t\psi(\alpha) = \lim_n t_n\psi(\alpha) = \lim_n \psi(c_{t_n}\alpha) = \psi(d\alpha)$. Hence $\lim c_{t_n} = d = c_t$. One shows easily that $c_t = t^{1/c}\, c$. Therefore, if $\alpha > 0$,

$$\psi(\alpha \cdot 1) = \psi((\alpha^c)^{1/c} \cdot 1) = \alpha^c\psi(1),$$

and if $\alpha < 0$,

$$\psi(\alpha) = \psi(|\alpha|(-1)) = |\alpha|^c\psi(-1) = |\alpha|^c\,\overline{\psi(1)};$$

for, from the form of $\psi(\alpha)$, we see that $\overline{\psi(\alpha)} = \psi(-\alpha)$. Thus, if $\psi(1) = -a_o + a_1 i$, we have $\psi(\alpha) = |\alpha|^c(-a_o + ia_1 \frac{\alpha}{|\alpha|})$. Since $|e^{\psi(\alpha)}| \leq 1$, $a_o \geqslant 0$; if $a_o = 0$, $E(e^{i\alpha x_1}) = e^{ia_1\alpha \cdot |\alpha|^{c-1}}$, so that

$$E[\cos(\alpha(x_1 - a_1|\alpha|^{c-1}))] = 1, \text{ i.e. } E[1 - \cos(\alpha(x_1 - a_1|\alpha|^{c-1}))] = 0.$$

We should therefore have $\cos(\alpha(x_1 - a_1|\alpha|^{c-1})) = 1$ a.e. or $\alpha[x_1(w) - a_1|\alpha|^{c-1}] = 2k(\alpha, w)\pi$, $k(\alpha, w)$ being an integer depending on α and w. For fixed w, thus $k(\alpha, w)$ is continuous in α. Letting $\alpha \to 0$, we see that $k(\alpha, w) \equiv 0$. Therefore $x_1(w) - a_1|\alpha|^{c-1} \equiv 0$. If $a_1 \neq 0$, this shows that $c = 1$, so that $\psi(\alpha) = ia_1\alpha$.

We shall now show that $0 < c \leq 2$. We have $x_t \underset{L}{\sim} t^{1/c} x_1$, $x_{st} \underset{L}{\sim} (st)^{1/c} x_1 \underset{L}{\sim} s^{1/c} x_t$. By using additivity and homogeneity of x_t and x_{st}, we can show that $x_{s\cdot} \underset{L}{\sim} s^{1/c} x_{\cdot}$ (as random processes). It follows that the expectations of the number of jumps of these processes are the same (because, if $p_1(E_t)$ and $p_2(E_t)$ correspond to $x_{s\cdot}$ and $s^{1/c} x_{\cdot}$, then $p_1(E_t)$, $p_2(E_t)$ are equivalent in law). The expected number of jumps of $x_{s\cdot}$ and $s^{1/c} x_{\cdot}$ in $dt\,du$ are $s\,dt\,\sigma(du)$ and $dt\,\sigma(s^{-1/c}du)$ respectively. We have therefore $\sigma(s^{-1/c}du) = s\,\sigma(du)$. Let $\sigma_+(u) = \int_u^\infty \sigma(du)$ for $u > 0$. Then since $s\,\sigma(du) = \sigma(s^{-1/c}du)$,

$$s\,\sigma_+(u) = s\int_u^\infty \sigma(du) = \int_u^\infty \sigma(s^{-1/c}du) = \int_{s^{-1/c}u}^\infty \sigma(du) = \sigma_+(us^{-1/c}).$$

Putting $s = u^c$ and $a_+ = c\,\sigma_+(1)(\geq 0)$, we get $u^c\sigma_+(u) = \sigma_+(1) = \frac{a_+}{c}$, so that $\sigma_+(u) = \frac{a_+}{c} u^{-c}$. Therefore $\sigma(du) = a_+ u^{-c-1}\,du$. Similarly, we see that $\sigma(du) = a_-|u|^{-c-1}$ $(u < 0)$. If $a_+ = a_- = 0$, then $\psi(\alpha) = i\,m\alpha - v/2\,\alpha^2$ and x_t is Gaussian additive. Also, $\psi(\alpha) = |\alpha|^c(-a_0 + i\,a_1 \frac{\alpha}{|\alpha|})$ so that $c = 2$, $v/2 = a_0$ and $a_1 = m = 0$. Therefore $\psi(\alpha) = -a_0|\alpha|^2$, $a_0 > 0$.

Let us now assume that at least one of a_+ or a_- is positive, say a_+. Since $\int_{-1}^1 u^2\sigma(du) < \infty$, $\int_0^1 u^2\sigma(du) < \infty$ so that $a_+\int_0^1 u^2 \frac{du}{u^{c+1}} < \infty$. This proves that $c < 2$. Again, using $\int_1^\infty \sigma(du) < \infty$, we can see that $0 < c$. The theorem is completely proved.

The number c is called the <u>index</u> of the stable process. We shall discuss the cases $o < c < 1$, $c = 1$, and $1 < c < 2$.

<u>Case (a)</u> $o < c < 1$.

In this case we have $\int_{-\infty}^{\infty} \sigma(du) = \infty$, $\int_{-1}^{1} |u| \sigma(du) < \infty$. The second inequality implies $E(\int_{-1}^{1} |u|\ p([o,t] \times du)) < \infty$ so that

$$P(\int_{-1}^{1} |u|\, p([o,t] \times du) < \infty) = 1.$$

Let $f(n) = t \int_{|u| \geq \frac{1}{n}} \sigma(du) = \int_{|u| \geq \frac{1}{n}} \sigma([o,t] \times du) = E(\int_{|u| \geq \frac{1}{n}} p([o,t] \times du))$

$$= E(p([o,t] \times (|u| \geq \tfrac{1}{n}))).$$

Since $p(E_t)$ is a Poisson variable we have

$$P\left[p([o,t] \times (|u| \geq \tfrac{1}{n})) \geq N\right] = \sum_{k \geq N} e^{-f(n)} \frac{[f(n)]^k}{k!} = 1 - e^{-f(n)} \sum_{k < N} \frac{[f(n)]^k}{k!}$$

Letting $n \longrightarrow \infty$, since $f(n) \longrightarrow \infty$ we have

$$P\left[p([o,t] \times (|u| > o)) \geq N\right] = 1.$$

Hence $$P\left[\text{the number of jumps in } [o,t] = \infty\right] = 1 \cdot$$

Now $\int_{-\infty}^{\infty} \frac{|u|}{1+u^2} \sigma(du) < \infty$, so that we oan write

$$x_t = g_2(t) + \int_{-\infty}^{\infty} u\ p([o,t] \times du) .$$

We can now show that

$$\psi(\alpha) = i\, m\alpha - \frac{v}{2}\alpha^2 + a_+ \int_{o}^{\infty} [e^{i\alpha u} - 1] \frac{du}{u^{c+1}} + a_- \int_{-\infty}^{o} [e^{i\alpha u} - 1] \frac{du}{|u|^{c+1}}$$

Also $\int_0^\infty (e^{i\alpha u}-1)\frac{du}{u^{c+1}} = \alpha^c\int_0^\infty [e^{iu}-1]\frac{du}{u^{c+1}} = O(|\alpha|^c) = o(|\alpha|) = o(|\alpha|^2)$;

similarly $\int_{-\infty}^0 (e^{i\alpha u}-1)\frac{du}{|u|^{c+1}} = O(|\alpha|^c) = o(|\alpha|) = o(|\alpha|^2)$ as $\alpha \longrightarrow \infty$. Hence $v = m = 0$, and $\psi(\alpha) = a_+\int_0^\infty [e^{i\alpha u}-1]\frac{du}{u^{c+1}} + a_-\int_{-\infty}^0 [e^{i\alpha u}-1]\frac{du}{|u|^{c+1}}$ and $x_t = \int_{-\infty}^\infty u\, p([o,t]\times du)$.

<u>Case (b)</u> $1 < c < 2$.

In this case $\int_{-\infty}^\infty \frac{u^2}{1+|u|}\frac{du}{|u|^{c+1}} < \infty$. Hence we can write

$$\psi(\alpha) = im\alpha - \frac{v}{2}\alpha^2 + a_+\int_0^\infty \left[e^{i\alpha u}-1-i\alpha u\right]\frac{du}{u^{c+1}} + a_-\int_{-\infty}^0 \left[e^{i\alpha u}-1-i\alpha u\right]\frac{du}{|u|^{c+1}}$$

Now $\psi(\alpha) = O(|\alpha|^c)$, so that comparing the orders as $\alpha \longrightarrow \infty$ and $\alpha \longrightarrow o$ we see immediately that $m = v = 0$. Hence

$$\psi(\alpha) = a_+\int_0^\infty \left[e^{i\alpha u}-1-i\alpha u\right]\frac{du}{u^{c+1}} + a_-\int_{-\infty}^0 \left[e^{i\alpha u}-1-i\alpha u\right]\frac{du}{|u|^{c+1}}.$$

We have

$$E\left(\int_{-1}^1 |u|^{c'}\, p([o,t]\times du)\right) = a_+\, t\int_0^1 |u|^{c'}\frac{du}{u^{c+1}} + a_-\, t\int_{-1}^0 |u|^{c'}\frac{du}{|u|^{c+1}},$$

which is finite or infinite according as $c' > c$ or $c' \leq c$. Therefore $P\left[\sum_{s\leq t} |x_s - x_{s-}|^{c'} < \infty\right] = 1$ if $c' > c$. We can easily show that

$$E\left(\exp\left(-\int_{-\delta}^{\delta} |u|^{c'}\, p([o,t]\times du)\right)\right) = \exp\left(-t\int_{-\delta}^{\delta}(1-e^{-|u|^{c'}})\,\sigma(du)\right).$$

Since $\int_{-1}^1 |u|^{c'}\sigma(du) = \infty$ the right side is zero in the limit. It follows that $\exp\left(-\int_{-1}^1 |u|^{c'}\, p([o,t]\times du)\right) = 0$ with probability 1.

Hence $\quad P\left[\sum_{s\leq t}|x_s - x_{s-}|^{c'} = \infty\right] = 1$, if $c' \leq c$.

<u>Case (c)</u> $\quad c = 1$.

We have $\psi(\alpha) = i\, a_1\alpha - a_0|\alpha|$. Since

$$-\pi|\alpha| = 2\int_0^\infty [\cos\alpha u - 1]\frac{du}{u^2} = \int_{-\infty}^{\infty}\left[e^{i\alpha u} - 1 - \frac{i\alpha u}{1+u^2}\right]\frac{du}{u^2} \quad ,$$

we have

$$\psi(\alpha) = i\, a_1\alpha + \frac{a_0}{\pi}\int_{-\infty}^{\infty}\left[e^{i\alpha u} - 1 - \frac{i\alpha u}{1+u^2}\right]\frac{du}{u^2} = im\alpha - \frac{v}{2}\alpha^2 + \int_{\infty}^{\infty}\left(e^{i\alpha u} - 1 - \frac{i\alpha u}{1+u^2}\right)\sigma(du).$$

From the uniqueness of representation of $\psi(\alpha)$ we get $v = 0$ and $\sigma(du) = \frac{du}{u^2}$. In this case thus $a_+ = a_-$ and

$$\psi(\alpha) = i\, a_1\alpha + \frac{a_0}{\pi}\int_{-\infty}^{\infty}\left[e^{i\alpha u} - 1 - \frac{i\alpha u}{1+u^2}\right]\frac{du}{u^2} \quad .$$

<u>Definition</u>. Processes for which $c = 1$ are called <u>Cauchy processes</u>.

§6. <u>Lévy process as a Markov process.</u>

Let $(x_t(w))$, $w \in \Omega(\mathcal{B}, P)$ be a temporally homogeneous Lévy process. Let $\mathcal{M} = (R^1, W, P_a)$, where $W = W_{d_1}$ and $P_a(B) = P(x_\cdot + a \in B)$. We show that $\mathcal{M}$ is Markov process.

If x is a random variable on a probability space Ω, then the map $(w,a) \longrightarrow (x(w),a)$ is measurable. It follows that the map $(w,a) \longrightarrow x(w) + a$ is measurable in the pair (w,a). Now note that if F is a fixed subset of $\Omega \times R^1$, then $f(a) = P(w:(w,a)\in F)$ is measurable in a. Hence $P(w:x(w)+a\in E)$ for $E \in \mathcal{B}(R^1)$ is measurable in a.

Therefore $P(t,a,E) = P(w:x_t(w)+a \in E)$ is measurable in a. If U is an open set containing a , U-a is an open set containing 0. Since x_t is continuous in probability

$$\lim_{t\to 0} P(t,a,U) = \lim_{t\to 0} P\left[x_t(w)\in U-a\right] = 1.$$

It remains to prove that if $t_1 < \ldots < t_n$,

$$P_a(x_{t_i}\in E_i, 1\le i\le n) = \int\cdots\int_{a_i\in E_i} P(t_1,a,da_1)P(t_2-t_1,a_1,da_2)\ldots P(t_n-t_{n-1},a_{n-1},da_n)$$

We prove this for n=2. We have, since $x_{t_2}-x_{t_1} \underset{L}{\sim} x_{t_2-t_1}$,

$$\int_{a_1\in E_1}\int_{a_2\in E_2} P(t_1,a,da_1)P(t_2-t_1,a_1,da_2) = \int_{a_1\in E_1} P(t_1,a,da_1)P(t_2-t_1,a_1,E_2)$$

$$= \int_{a_1\in E_1} P(x_{t_1}\in da_1-a)P(x_{t_2-t_1}\in E_2-a_1)$$

$$= \int_{a_1\in E_1} P(x_{t_1}\in da_1-a)P(x_{t_2}-x_{t_1}\in E_2-a-(a_1-a))$$

$$= P\left[(x_{t_1},x_{t_2}-x_{t_1})\in (E_2-a)' \cap ((E_1-a)\times R^1)\right]$$

$$= P\left[x_{t_1}\in E_1-a, x_{t_2}\in E_2-a\right] = P_a\left[x_{t_1}\in E_1, x_{t_2}\in E_2\right]$$

where $(E_2-a)' = \left\{(\xi,\eta) : (\xi,\eta)\in R^2 \text{ and } \xi+\eta\in E_2-a\right\}$.

Thus $\mathcal{M}$ is a Markov process. Further since $H_t f(a) = f(b)\, P(t,a,db) = \int f(a+b)\, P(x_t\in db)$, we see that $H_t(C(R^1))\subset C(R^1)$. $\mathcal{M}$ is thus strongly Markov. $\mathcal{M}$ is conservative. Recall that W_{d_1} consists of all functions which are of d_1-type before their killing time. We have

$$P_a(\sigma_\infty = \infty) = P_a(w: w(n) \in R^1 \text{ for every integer } n)$$

$$= \lim_n P_a(w(n) \in R^1) = \lim_n P(w: x_n(w) \in R^1) = 1.$$

Also $\mathcal{M}$ is translation invariant, i.e. if $\tau_h b = b + h$ then $P_{\tau_h a}(\tau_h B) = P_a(B)$.

Conversely any conservative translation invariant Markov process with state space R^1 can be got in the above way from a temporally homogeneous Lévy process.

We shall now prove that the kernel of G_α is the set of functions which are zero a.e., i.e. $G_\alpha f = 0$ implies $f = 0$ a.e. To prove this firstly observe that $G_\alpha f = 0$ implies $H_t f(a) = 0$ for almost all t. Hence we can find a sequence of $t_n \downarrow 0$ such that $H_{t_n} f(a) = 0$. Now $\int f(a+b)\,\varphi(t_n, db) = 0$. Since f is bounded it is locally summable. Hence for any interval (α, β) we have

$$\int\int_\alpha^\beta f(a+b)\,\varphi(t_n, db)\,da = 0 \text{ i.e.}$$

$$0 = \int\int_\alpha^\beta f(a+b)da\,\varphi(t_n, db) = \int\int_{\alpha+b}^{\beta+b} f(a)da\,\varphi(t_n, db)$$

$$= \int g(b)\,\varphi(t_n, db) = 0$$

where $g(b) = \int_{\alpha+b}^{\beta+b} f(a)da$ is continuous. It follows that $\int g(b)\,\varphi(t_n, db) \longrightarrow g(0)$ as $t_n \longrightarrow 0$ i.e. $\int_\alpha^\beta f(a)da = 0$. Since this is true for every interval (α, β), $f = 0$ a.e. This proves our contention.

<u>Generator</u>. It is difficult to determine the generator of this process in the general case. However we will determine $\mathcal{G}u$ when u satisfies some conditions.

<u>Theorem 1</u>

Let $\hat{f}(\eta) = \int e^{-i\eta a} f(a)da$ denote the Fourier transform of f. If $u = G_\alpha h$ with $h \in L^1(-\infty,\infty)$, then $u \in L^1$ and $\hat{u} = \frac{\hat{h}}{\alpha - \psi}$. Therefore $\mathcal{G}u \in L^1$ and $\widehat{\mathcal{G}u} = \psi\hat{u}$.

<u>Proof</u>. Let $\varphi(t,E) = P(x_t \in E)$. Then if $f \geqslant 0$ we have

$$\int H_t f(a)da = \int da \int f(a+b)\varphi(t,db) = \int \varphi(t,db) \int f(a+b)da$$

$$= \int \varphi(t,db) \int f(a)da = \int f(a)da$$

so that if $f \in L^1$, so is $H_t f(a)$. Now $H_t f(a)$ is measurable in the pair (t,a). We have similarly if $f \geqslant 0$,

$$\int G_\alpha f(a)da = \int da \int_0^\infty e^{-\alpha t} H_t f(a)dt = \int_0^\infty e^{-\alpha t} dt \int H_t f(a)da = \int_0^\infty e^{-\alpha t} dt \int f(a)da = \frac{1}{\alpha}\int f(a)da$$

so that $G_\alpha f(a) \in L^1$. Therefore

$$\widehat{G_\alpha h}(\eta) = \int e^{-ia\eta} da \int_0^\infty e^{-\alpha t} dt \int h(a+c)\,\varphi(t,dc).$$

Since

$$\left|\iiint e^{-\alpha t} h(a+c) e^{-ia\eta} da\, dt\, \varphi(t,dc)\right| \leq \iiint e^{-\alpha t} |h(a+c)|\, da\, dt\, \varphi(t,dc)$$

$$= \int G_\alpha |h(a)|\, da = \frac{1}{\alpha}\int |h(a)|\, da$$

we can interchange the orders of integration as we like. We have

$$\widehat{G_\alpha h}(\eta) = \int_0^\infty e^{-\alpha t} \hat{h}(\eta) \int e^{i\eta c} \varphi(t,dc)dt = \int_0^\infty e^{-\alpha t} \hat{h}(\eta) e^{t\psi(\eta)} dt = \frac{\hat{h}(\eta)}{\alpha - \psi(\eta)}$$

since $\int e^{i\alpha a} \varphi(t,da) = E(e^{i\alpha x_t}) = e^{t\psi(\alpha)}$ and since the real part of $\psi(\alpha)$ is non-positive $\int_0^\infty e^{-(\alpha - \psi(\eta))t} dt$ exists and equals $\frac{1}{\alpha - \psi(\eta)}$. Since $u = G_\alpha h$ is in L^1, $\mathcal{G}u \in L^1$. Also from the last equation $\alpha \widehat{G_\alpha h} - \hat{h} = \psi \widehat{G_\alpha h}$ so that $\widehat{\mathcal{G}u} = \psi \hat{u}$.

<u>Corollary</u>. If $\alpha > 0$ and $(\alpha - \psi)\hat{u} = \hat{f}$ for some function $f \in L^1$, then $u = G_\alpha f \in \mathcal{D}(\mathcal{G})$ and $\widehat{\mathcal{G}u} = \psi \hat{u}$.

For we have from Theorem 1, $\widehat{G_\alpha f} = \frac{\hat{f}}{\alpha - \psi} = \hat{u}$ so that $u = G_\alpha f$ (a.e) and $\widehat{\mathcal{G}u} = \psi \hat{u}$.

<u>Theorem 2</u>. If u, u' and u'' are in L^1, then $u \in \mathcal{D}(\mathcal{G})$ and u is given a.e. by

$$\mathcal{G}u(a) = m\,u'(a) + \frac{v}{2} u''(a) + \int_{-\infty}^{\infty} \left[u(a+b) - u(a) - \frac{b\,u'(a)}{1+b^2}\right] \sigma(db).$$

<u>Proof</u>. Let $f_1 = m\,u'$, $f_2 = \frac{v}{2} u''$, $f_3 = \int_{|b|>1} \left[u(a+b) - u(a) - \frac{b\,u'(a)}{1+b^2}\right] \sigma(db)$,

$f_4 = \int_{|b| \leq 1} \frac{b^3}{1+b^2} u'(a) \sigma(db)$ and $f_5 = \int_{|b| \leq 1} [u(a+b) - u(a) - b\,u'(a)] \sigma(db)$.

From the hypothesis we see that $f_i \in L^1$, $i = 1,2,3,4$. We prove that f_5 exists and is in L^1. We have

$$u(a+b) - u(a) - b\,u'(a) = \int_0^b u'(a+x)dx - b\,u'(a) = \int_0^b [u'(a+x) - u'(a)]\,dx$$

$$= \int_{x=0}^{b} dx \int_{y=0}^{x} u''(a+y)dy.$$

Therefore

$$\int_{-\infty}^{\infty} da \int_{|b|\leq 1} |u(a+b)-u(a)-bu'(a)|\,\sigma(db) \leq \int_{-\infty}^{\infty} da \int_{|b|\leq 1} \sigma(db) \int_{x=0}^{b} dx \int_{y=0}^{x} |u''(a+y)|\,dy$$

$$= \int_{|b|\leq 1} \sigma(db) \int_{x=0}^{b} dx \int_{y=0}^{x} dy \int_{-\infty}^{\infty} |u''(a+y)|\,da$$

$$= \int_{|b|\leq 1} \sigma(db) \int_{x=0}^{b} dx \int_{y=0}^{x} dy\,\|u''\| = \frac{\|u''\|_1}{2} \int_{|b|\leq 1} b^2 \sigma(db) < \infty .$$

This shows that f_5 exists, is in L^1 and $\| f_5 \|_1 \leq \frac{\|u''\|_1}{2} \int_{|b|\leq 1} b^2 \sigma(db)$.

We can easily see that

$$\hat{f}_1(\eta) = im\eta\hat{u}(\eta), \hat{f}_2(\eta) = -\frac{v}{2}\eta^2\hat{u}(\eta), f_3(\eta) = \hat{u}(\eta) \int_{|b|>1} \left[e^{i\eta b} - 1 - \frac{i\eta b}{1+b^2}\right] \sigma(db)$$

and $\hat{f}_4(\eta) = i\eta\hat{u}(\eta) \int_{|b|\leq 1} \frac{b^3}{1+b^2} \sigma(db)$. Further $f_5 \in L^1$ and we have seen that $\iint |u(a+b)-u(a)-bu'(a)|\,\sigma(db)da$ exists as a double integral. Hence we can interchange the order of integration in

$$\iint e^{-i\eta a} [u(a+b)-u(a)-bu'(a)]\,\sigma(db)da .$$

Thus we have $\hat{f}_5(\eta) = \int [e^{i\eta b} - 1 - i\eta b]\,\hat{u}(\eta)\,\sigma(db)$. Hence finally if $f = f_1 + \ldots + f_5$, $\hat{f}(\eta) = \hat{f}_1(\eta) + \ldots + \hat{f}_5(\eta) = \psi(\eta)\,\hat{u}(\eta)$. We have $[\alpha - \psi(\eta)]\hat{u}(\eta) = \alpha\hat{u} - \hat{f} = \widehat{\alpha u - f}$. Using the corollary of Theorem 1, we see that $u \in \mathcal{D}(\mathcal{G})$ and $u = G_\alpha[\alpha u - f]$ so that $\mathcal{G}u = \alpha u - (\alpha u - f) = f$(a.e). This proves the theorem.

Remark. If $\varphi(t,E)$ is symmetric, $\mathcal{G}u(a) = \frac{v}{2} u''(a) + \int_0^\infty [u(a+b) + u(a-b) - 2u(a)]\,\sigma(db)$. In the case of a symmetric Cauchy process $v=0$ and $\mathcal{G}u(a) = \int_0^\infty [u(a+b)+u(a-b)-2u(a)]\,\sigma(db)$.

§7. Multidimensional Lévy processes.

A k-dimensional stochastic process (x_t) is called a k-dimensional Lévy process if, it is additive, almost all sample functions are d_1 and it has no point of fixed discontinuity; note that unlike the k-dimensional Brownian motion the component process need not be independent.

A k-dimensional random variable x is called Gaussian if and only if $E(e^{i(\alpha,x)}) = e^{i(m,\alpha)-\frac{1}{2}(v\alpha,\alpha)}$ where m is a vector, v a positive definite matrix and (a,b) denotes the scalar product of a and b.

Let $x=(x^1,\ldots,x^k)$ be a k-dimensional random variable such that for any real $c_1,\ldots,c_k$, $\Sigma c_j x^j$ is a Gaussian variable. Then x is also Gaussian. For $E(e^{i\beta\Sigma\alpha_j x^j}) = e^{im\beta-\frac{v'}{2}\beta^2}$ where $m = \Sigma\alpha_j m^j$, $v' = E((\Sigma\alpha_j(x^j-m^j))^2)$ with $m^j = E(x^j)$. Now $v' = \Sigma\alpha_i^2 v_{ii} + 2\sum_{i<j}\alpha_i\alpha_j v_{ij} = (v\alpha,\alpha)$ where $v_{ij} = E((x^i-m^i)(x^j-m^j))$ and $v = (v_{ij})$. Since $v' \geq 0$, v is a positive definite matrix. Putting $\beta = 1$ we have $E(e^{i(\alpha,x)}) = E(e^{i\Sigma\alpha_j x^j}) = e^{i(m,\alpha)-\frac{1}{2}(v\alpha,\alpha)}$.

Thus if almost all sample functions of a k-dimensional Lévy process (x_t) are continuous then x_t-x_s is Gaussian.

Let $(x_t(w))$ be a k-dimensional Lévy process. Proceeding exactly as in the case of $k = 1$, we can show that

$$x_t = g(t) + \int_{R^k\times[o,t]}\left[up(ds\,du) - \frac{u}{1+u^2}\,\sigma(ds\,du)\right].$$

where $g(t)$ is continuous; hence we can obtain

$$\log E(e^{i(\alpha,x_t)}) = i(m(t),\alpha)-\tfrac{1}{2}(v(t)\alpha,\alpha)+\int_{R^k\times[o,t]}\left[e^{i(\alpha,b)}-1-\frac{i(\alpha,b)}{1+|b|^2}\right]\sigma(ds\,db)$$

If $\sigma = 0$ the path functions are continuous.

If (x_t) is rotation invariant i.e. if $E(e^{i(\alpha,x_t)}) = E(e^{i(\alpha,\theta x_t)})$ where θ is any rotation, we have, since $(\alpha,\theta^{-1}x_t) = (\theta\alpha,x_t)$ $(m(t),\theta\alpha) = (m(t),\alpha)$ and $(v(t)\theta\alpha,\theta\alpha) = (v(t)\alpha,\alpha)$. Since this is true for every **rotation** θ we should have $m(t) \equiv 0$ and $v(t)$ a diagonal matrix in which all the diagonal elements are the same and we can write

$$\log E(e^{i(\alpha,x_t)}) = -\tfrac{1}{2}v(t)|\alpha|^2+\int_{[o,t]\times R^k}\left[e^{i(\alpha,b)}-1-\frac{i(\alpha,b)}{1+b^2}\right]\sigma(ds\,db)\ .$$

If the process is temporally homogeneous, $E(e^{i(\alpha,x_t)}) = e^{t\psi(\alpha)}$ where $\psi(\alpha) = i(m,\alpha)-\tfrac{1}{2}(v\alpha,\alpha)+\int_{R^k}\left[e^{i(\alpha,b)}-1-\frac{i(\alpha,b)}{1+b^2}\right]\sigma(db)$.

Now suppose that (x_t) is a stable process i.e. (x_t) is temporally homogeneous and $x_t \underset{L}{\sim} c_t\, x_1$. We can show (proceeding in the same way as for $k=1$) that $\sigma(a\,E) = \frac{1}{a^c}\,\sigma(E)$ for $a > o$. Now we prove that $o < c < 2$ unless $\sigma \equiv o$. Let $E = (b: 1 \geq |b| > \tfrac{1}{2})$. Since $\int_{|b|\leq 1} |b|^2\,\sigma(db) < \infty$, we have

$$\sum_{n=o}^{\infty}\int_{\frac{1}{2^n}\geq|b|\geq\frac{1}{2}\cdot\frac{1}{2^n}}|b|^2\,\sigma(db) < \infty\ .$$

so that

$$\sum_{n=o}^{\infty}\frac{1}{2^2}\cdot\frac{1}{2^{2n}}\sigma\left(b : \frac{1}{2^n}\geq|b|>\frac{1}{2}\,\frac{1}{2^n}\right) < \infty \text{ i.e. } \sum\frac{1}{2^{2n}}\sigma\left(\frac{1}{2^n}\,E\right) < \infty\ .$$

Hence since $\sigma(r\,E) = \frac{1}{r^c}\,\sigma(E)$, we should have $\sigma(E)\sum \frac{2^{nc}}{2^{2n}} < \infty$. If $\sigma(E) \neq 0$, $c < 2$. Similarly considering $\int_{|b| \geq 1} \sigma(db) < \infty$ we can prove that $c > 0$.

Let S denote the surface of the unit sphere in R^k. Then R^k minus the point $(0,0,\ldots,0)$ can be regarded as the product of S and the half line $(0,\infty)$. For any Borel subset Θ of S let $c^{-1}\sigma_+(\Theta) = \sigma(\Theta \times [1,\infty))$. Then $c^{-1}\sigma_+(d\Theta)$ is a measure on $\mathcal{B}(S)$ and

$$\sigma(\Theta \times [r,\infty)) = \sigma(r.\,\Theta \times [1,\infty)) = \frac{1}{r^c}\,c^{-1}\sigma_+(\Theta) = \frac{1}{c}\int_{[r,\infty)\times\Theta} \frac{dr}{r^{c+1}}\,c\,\sigma_+(d\Theta)$$

It follows that $\sigma(db) = \frac{dr}{r^{c+1}}\,\sigma_+(d\Theta)$.

If x_t is rotation invariant $\sigma_+(d\Theta)$ will be rotation invariant and hence must be the uniform distribution $d\Theta$ so that $\sigma(db) = \text{const.}\,\frac{dr\,d\Theta}{r^{c+1}}$.

We can consider a k-dimensional temporally homogenous Lévy process (x_t) as a Markov process with state space R^k and we can prove that if $f \in L^1(R^k)$ and $u = G_\alpha f$ then $(\alpha - \psi(\xi))\,\hat{u}(\xi) = \hat{f}(\xi)$ and $\widehat{\mathcal{G}u}(\xi) = \psi(\xi)\,\hat{u}(\xi)$. If $u \in L^1$ and $(\alpha - \psi(\xi))\,\hat{u}(\xi) \in L^1$ then $u \in \mathcal{D}(\mathcal{G})$ and $\widehat{\mathcal{G}u}(\xi) = \psi(\xi)\,\hat{u}(\xi)$. To prove this let $\hat{f} = (\alpha - \psi(\xi))\,\hat{u}(\xi)$ and $v = G_\alpha f$. Then $(\alpha - \psi(\xi))\,\hat{v} = \hat{f} = (\alpha - \psi(\xi))\,\hat{u}$ so that $u = v$ a.e. and $u \in \mathcal{D}(\mathcal{G})$.

Now suppose that (x_t) is stable and rotation invariant. We can show that $\psi(\alpha) = |\alpha|^c\,\psi(\frac{\alpha}{|\alpha|})$ so that if $\psi(\alpha)$ is rotation invariant $\psi(\alpha) = -|\alpha|^c \cdot$ constant. If we look at the expression for $\psi(\alpha)$, we see that real part of $\psi(\alpha) \leq 0$. It follows that const. ≥ 0. In this case, we thus have

$$\widehat{\mathcal{G}u}(\xi) = -\lambda\, \hat{u}(\xi)\, |\xi|^c \text{ i.e. } \hat{u}(\xi) = -\frac{1}{\lambda|\xi|^c}\widehat{\mathcal{G}u}(\xi).$$

The Fourier transform (in the distribution sense) of $|a|^{c-k}$ is $\mu|\xi|^{-c}$, $\mu = \pi^{(k/2)-c}\Gamma(c/2)/\Gamma(\frac{k-c}{2})$ (refer to Théorie des distributions by Schwartz, page 113, Example 5). Since $\mathcal{G}u$ is bounded, it is a rapidly decreasing distribution. Hence (see page 124, Théorie des distributions, Schwartz)

$$\hat{u}(\xi) = A\,\widehat{\mathcal{G}u}(\xi)\,\widehat{\frac{1}{|a|^{k-c}}}(\xi) = A\,\widehat{\mathcal{G}u * \frac{1}{|a|^{k-c}}}(\xi),\ A = -\frac{1}{\mu\lambda}.$$

Therefore $u(a) = A\int \mathcal{G}u(b)\,\frac{1}{|a-b|^{k-c}}\,db$. Thus $\frac{1}{|a-b|^{k-c}}$ is the potential kernel corresponding to this process. Potentials with such kernels are called <u>Riesz Potentials.</u>

When $c = 2$, $u(a) = A\,\mathcal{G}u * \frac{1}{|a|^{k-2}}$ so that

$$\Delta u(a) = A\,\mathcal{G}u * \Delta\frac{1}{|a|^{k-2}} = A\,\mathcal{G}u(a).$$

Section 5

Stochastic differential equations

§1. Introduction.

The standard Brownian motion is a one-dimensional diffusion whose generator is $\frac{1}{2}\dfrac{d^2}{da^2}$. We shall here construct a more general one-dimensional diffusion whose generator $\mathcal{G}$ is the differential operator

$$D = \tfrac{1}{2}\, p^2(a)\, \frac{d^2}{da^2} + r(a)\, \frac{d}{da} \; ;$$

precisely if $u \in C_2(R^1) = \{ u : u, u', u'' \text{ continuous and bounded} \}$ then $u \in \mathcal{D}(\mathcal{G})$ and $\mathcal{G}\, u = Du$. To do this we consider the stochastic differential equation

$$dx_t = p(x_t)\, d\beta_t + r(x_t)\, dt ,$$

where β_t is a Wiener process. The meaning of the above equation is

$$x_u - x_t = \int_t^u p(x_s) d\beta_s + \int_t^u r(x_s) ds \quad , \; 0 \leq t < u < \infty .$$

The meaning of $\int_t^u p(x_s) d\beta_s$ has to be made clear; we do this in article 3. Note that it cannot be interpreted as a Stieltjes integral for a fixed path because it can be shown that as a function of s, β_s is not of bounded variation for almost all paths. We make the following formal considerations postponing the definition of the integral to §3.

Let $x_t^{(a)}$ be a solution of the differential equation with the initial condition $x_o^{(a)} = a$, i.e. let $x_t^{(a)}$ be a solution of the integral equation

$$x_t = a + \int_o^t p(x_s)d\beta_s + \int_o^t r(x_s)ds .$$

Then, under certain regularity conditions on p and r , we can define a strong Markov process $\mathcal{M} = (S,W,P_a)$ with $S = R^1$, $W = W_c(R^1)$, $P_a(B) = P(x_\cdot^{(a)} \in B)$ and such that

$$\mathcal{G}u(a) = \tfrac{1}{2}\, p^2(a) \frac{d^2u}{da^2} + r(a)\frac{du}{da} \quad , \quad u \in C_2(R^1)$$

where $\mathcal{G}$ is the generator in the restricted sense. The same can be done in multi-dimensional case replacing β ,p,r by a multi-dimensional Wiener process, a matrix valued function and a vector valued function respectively. Componentwise,we will have

$$dx_t^i = \sum_j p_j^i(x_t)d\beta_t^j + r^i(x_t)dt \;,\quad i=1,\ldots,n$$

and the generator will be given by

$$\mathcal{G}u(a) = \tfrac{1}{2}\sum_{i,j} q^{ij}(a)\frac{\partial^2 u(a)}{\partial a^i \partial a^j} + \sum_i r^i(a)\frac{\partial u(a)}{\partial a^i}.$$

where $q^{ij} = \sum_k p_k^i p_k^j$.

Taking local coordinates we can extend the above to the case in which the state space S is a manifold.

Coming back to stochastic integrals we prove the following theorem which shows that $\int_t^u f(s,w)d\beta(s,w)$ cannot be interpreted as a Stieltjes integral.

<u>Theorem</u>. Let Δ be the subdivision $t = s_o < s_1 < \ldots < s_n = u$, and $\delta(\Delta) = \max_i (s_{i+1} - s_i)$. Then

1. $r_2(\Delta) = \sum_i (\beta(s_{i+1})-\beta(s_i))^2 \longrightarrow u-t$ L^2-mean as $\delta(\Delta) \to 0$

2. $r(\beta, t, u) = \sup_\Delta \sum_i |\beta(s_{i+1})-\beta(s_i)| = \infty$ with probability 1.

<u>Proof</u>. 1) $E(r_2(\Delta)) = \sum_i E[\beta(s_{i+1})-\beta(s_i)]^2 = \sum_i (s_{i+1}-s_i) = u-t$ and

$$E(r_2(\Delta)^2) = \sum_i E((\beta(s_{i+1})-\beta(s_i))^4) + 2\sum_{i<j} E((\beta(s_{i+1})-\beta(s_i))^2(\beta(s_{j+1})-\beta(s_j))^2)$$

$$= \sum_i 3(s_{i+1}-s_i)^2 + 2\sum_{i<j} E((\beta(s_{i+1})-\beta(s_i))^2)E((\beta(s_{j+1})-\beta(s_j))^2)$$

$$= \sum_i 3(s_{i+1}-s_i)^2 + 2\sum_{i<j} (s_{i+1}-s_i)(s_{j+1}-s_j) = 2\sum_i (s_{i+1}-s_i)^2 + \sum_i (s_{i+1}-s_i)^2 + \sum_{i<j} 2(s_{i+1}-s_i)(s_{j+1}-s_j)$$

$$= 2\sum_i (s_{i+1}-s_i)^2 + \left[\sum(s_{i+1}-s_i)\right]^2 = 2\sum(s_{i+1}-s_i)^2 + (u-t)^2$$

because $\beta(s_{i+1}) - \beta(s_i)$ and $\beta(s_{j+1}) - \beta(s_j)$ are independent for $i \neq j$ and $E((\beta(t)-\beta(s))^4) = 3(t-s)^2$. We thus have

$$E((r_2(\Delta)-(u-t))^2) = E(r_2(\Delta)^2)-(u-t)^2 = 2\sum_i (s_{i+1}-s_i)^2$$

$$\leq 2\delta(\Delta)\sum_i (s_{i+1}-s_i) \longrightarrow 0 \text{ as } \delta(\Delta) \longrightarrow 0.$$

2) From 1) we can find a sequence $\Delta^n = (t = s_o^{(n)} < \ldots < s_{p_n}^{(n)} = u)$ such that $r_2(\Delta^n) \longrightarrow u - t$ with probability 1. We have

$$r(\beta,t,u) \geqslant \Sigma|\beta(s_{i+1}^{(n)})-\beta(s_i^{(n)})| \geqslant \frac{\Sigma|\beta(s_{i+1}^{(n)})-\beta(s_i^{(n)})|^2}{\max_i|\beta(s_{i+1}^{(n)})-\beta(s_i^{(n)})|} \longrightarrow \infty$$

since $\Sigma|\beta(s_{i+1}^{(n)}) - \beta(s_i^{(n)})|^2 \longrightarrow u - t$ and $\max_i \Sigma|\beta(s_{i+1}^{(n)}) - \beta(s_i^{(n)})| \longrightarrow 0$ because of continuity of path functions.

§2. Stochastic integral (1) – Function spaces $\mathcal{E}, \mathcal{L}^2, \mathcal{L}_s$.

Let T be a time interval $[u,v)$, $0 \leq u < v < \infty$ and β_t, $t \in T$ be a Wiener process i.e. (1) the sample functions are continuous for almost all w, 2) $P(\beta_t-\beta_s \in E) = \int_E \frac{1}{\sqrt{2\pi(t-s)}} e^{-x^2/2(t-s)} dx$ and 3) $\beta_{t_1}, \beta_{t_2}-\beta_{t_1}, \ldots, \beta_{t_n}-\beta_{t_{n-1}}$ are independent if $t_1 < \ldots < t_n \in T$. Let $\mathcal{B}^t$, $t \in T$ be a monotone increasing system of Borel subalgebras of $\mathcal{B}$ such that $\mathcal{B}^t$ includes all null sets for each t, $\beta_t \in (\mathcal{B}^t)$ and $\beta_{t+h}-\beta_t$ is independent of $\mathcal{B}^t$ for $h > 0$. We shall use the notation $f \in (\mathcal{B})$ to denote that f is $\mathcal{B}$-measurable.

Let $\mathcal{L}_s$ be the set of all functions f such that 1) f is measurable in (t,w), 2) $f_t \in (\mathcal{B}^t)$ for almost all $t \in T$ and 3) $\int_T f_t^2 \, dt < \infty$ for almost all $w \in \Omega$. Instead of 3) we also consider the two stronger conditions

3') $\int_\Omega \int_T f(t,w)^2 dt \, dP < \infty$

3") there exist a subdivision $u = t_0 < t_1 < \ldots < t_n = v$ and $M < \infty$ such that

$$f_t(w) = f_{t_i}(w) \quad , \quad t_i \leq t < t_{i+1} \quad , \quad 0 \leq i \leq n-1$$

and $|f_t(w)| < M$.

We define the function spaces $\mathcal{L}^2$ and $\mathcal{E}$ by

$$\mathcal{L}^2 = \{f : 1), 2) \text{ and } 3') \text{ hold}\}$$

$$\mathcal{E} = \{f : 1), 2) \text{ and } 3'') \text{ hold}\}.$$

Clearly $\mathcal{E} \subset \mathcal{L}^2 \subset \mathcal{L}_s$. $\mathcal{L}^2$ is a (real) Hilbert space with the norm $\|f\|^2 = \int_\Omega \int_T |f|^2 \, dt \, dP$ and $\mathcal{L}_s$ is a (real) Fréchet space with the norm $\|f\|_{\mathcal{L}_s} = \int_\Omega \frac{1}{1+\sqrt{\int_T |f|^2 dt}} \cdot \sqrt{\int_T |f|^2} \, dt$. "$\|f\|_{\mathcal{L}_s} \to 0$" is equivalent to "$\int_T |f|^2 dt \to 0$ in probability" and if $f \in \mathcal{L}^2$ then $\|f\|_{\mathcal{L}_s} \leq \|f\|$.

<u>Theorem 1</u>. 1. $\mathcal{E}$ is dense in $\mathcal{L}^2$ (with the norm $\| \ \|$)

2. $\mathcal{E}$ is dense in $\mathcal{L}_s$ (with the norm $\| \ \|_{\mathcal{L}_s}$).

<u>Proof</u>. 1. We shall prove that, given $f \in \mathcal{L}^2$ there exists a sequence $f_n \in \mathcal{E}$ such that $\|f_n - f\| \longrightarrow 0$. We can assume that f is bounded. Put $f(t,w) = 0$ for $t \notin T$. Then f is defined for all t (this is to avoid changing T each time) and $\int_{-\infty}^{\infty} f^2 \, dt \, dP < \infty$ so that $\int_{-\infty}^{\infty} f^2 dt < \infty$ for almost all w. Therefore

$$\int_{-\infty}^{\infty} |f(t+h)-f(t)|^2 dt \longrightarrow 0 \text{ as } h \longrightarrow 0.$$

Also $\int_{-\infty}^{\infty} |f(t+h)-f(t)|^2 dt \leq 4 \int_{-\infty}^{\infty} f(t)^2 dt \in L^1(\Omega)$. We get

$$\int_\Omega \int_{-\infty}^{\infty} |f(t+h)-f(t)|^2 dt \, dP \longrightarrow 0 \text{ as } h \longrightarrow 0.$$

If $\varphi_n(t) = \frac{[2^n t]}{2^n}$, $n \geq 1$ then

$$\int_\Omega \int_{-\infty}^{\infty} |f(s+\varphi_n(t)) - f(s+t)|^2 ds\, dP \longrightarrow 0 \text{ as } n \longrightarrow \infty$$

Also $\int_\Omega \int_{-\infty}^{\infty} |f(s+\varphi_n(t)) - f(s+t)|^2 ds\, dP \leq 4 \int_\Omega \int_{-\infty}^{\infty} f(s)^2\, ds\, dP.$

Since $T = [u,v)$ is a finite interval

$$\int_{u-1}^{v} \int_\Omega \int_{-\infty}^{\infty} |f(s+\varphi_n(t)) - f(s+t)|^2 ds\, dP\, dt \longrightarrow 0 \text{ as } n \longrightarrow \infty$$

i.e. $\int_{-\infty}^{\infty} ds \int_{u-1}^{v} \int_\Omega |f(s+\varphi_n(t)) - f(s+t)|^2\, dP\, dt \longrightarrow 0$ as $n \longrightarrow \infty$.

Therefore there exists a subsequence $\{n_i\}$ such that

$$\int_{u-1}^{v} \int_\Omega |f(s+\varphi_{n_i}(t)) - f(s+t)|^2\, dP\, dt \longrightarrow 0 \text{ for almost all } s.$$

Choose $s \in [0,1]$ and fix it. Then

$$\int_{u-1}^{v} |f(s+\varphi_{n_i}(t)) - f(s+t)|^2\, dP\, dt \longrightarrow 0 \text{ as } n_i \longrightarrow \infty.$$

Changing the variable

$$\int_{u-1+s}^{v+s} \int_\Omega |f(s+\varphi_{n_i}(t-s)) - f(t)|^2\, dP\, dt \longrightarrow 0 \text{ as } n_i \longrightarrow \infty$$

since $0 \leq s \leq 1$

$$\int_{u}^{v} \int_\Omega |f(s+\varphi_{n_i}(t-s)) - f(t)|^2\, dP\, dt \longrightarrow 0.$$

Let $h_i(t) = f(s+\varphi_{n_i}(t-s))$. Then $h_i \in \mathcal{E}$ and $\| h_i - f \| \longrightarrow 0$.

2. Let $f \in \mathcal{L}_s$. We prove that there exists a sequence $f_n \in \mathcal{E}$ with $\| f_n - f \|_{\mathcal{L}_s} \longrightarrow 0$. We can assume that f is bounded so that $f \in \mathcal{L}^2$. We can find $f_n \in \mathcal{E}$ such that $\| f_n - f \| \longrightarrow 0$. But $\| f_n - f \|_{\mathcal{L}_s} \leq \int_\Omega \sqrt{\int_T | f_n - f |^2 dt}\, dP \leq \sqrt{\int_\Omega \int_T |f_n - f|^2 dt\, dP} = \| f_n - f \| \longrightarrow 0$.

<u>Remark</u>. Let f^M be the truncation of f by M i.e.

$$f^M = (f \vee (-M)) \wedge M$$

and for a subdivision $\Delta = (u = t_o < t_1 < \dots < t_n = v)$ let f_Δ be the function $f_\Delta(t,w) = f(t_i,w)$, $t_i \leq t < t_{i+1}$, $0 \leq i \leq n-1$. Then the approximating functions f_n in the above theorem are of the form $f_n = f_{\Delta_n}^{M_n}$ for some M_n, Δ_n.

§ 3. <u>Stochastic Integral</u> (2) - <u>Definitions and properties</u>.

Let $L^2(\Omega)$ be the real L^2-space with the usual L^2-norm $\| \;\|$ and $S(\Omega)$ be the space of all measurable functions with the norm $\| f \|_s = \int \frac{1}{1+|f(w)|} |f(w)|\, dP(w)$. $S(\Omega)$ is a real Fréchet space and " $\| f \|_s \longrightarrow 0$ " is equivalent to " $f \longrightarrow 0$ in probability". Clearly $L^2(\Omega) \subset S(\Omega)$ and if $f \in L^2(\Omega)$, $\| f \|_s \leq \| f \|$.

We first define $I(f) = \int_T f d\beta$ for $f \in \mathcal{E}$, show that it is continuous in the norms $\| \;\|$, $\| \;\|_s$ and hence that it is extendable to $\mathcal{L}^2$ and $\mathcal{L}_s$.

We define for $f \in \mathcal{E}$

$$I(f) = \int_T f_t\, d\beta_t = \sum_{i=o}^{n-1} f(t_i)(\beta(t_{i+1}) - \beta(t_i))$$

where $(t_0 = u < t_1 < \dots < t_n = v)$ is any subdivision by which f is expressed. This definition is independent of the division points with respect to which f is expressed and $I(f) \in L^2(\Omega) \subset S(\Omega)$. That I is linear is easy to see and

$$E(I(f)) = \sum_i E(f(t_i)(\beta(t_{i+1}) - \beta(t_i))) =$$

$$= \sum_i E(f(t_i))E(\beta(t_{i+1}) - \beta(t_i)) = 0$$

since $\beta(t_i)$ and $\beta(t_{i+1}) - \beta(t_i)$ are independent and $E(\beta(t)) = 0$.

Now we prove the following:

(A) $\| f \| = \| I(f) \|$. (Though we use the same notation, note that $f \in \mathcal{L}^2$, $I(f) \in L^2(\Omega)$).

(B) $\| I(f) \|_s = O(\| f \|_{\mathcal{L}_s}^{1/3})$.

<u>Proof of (A)</u>. Let $(f,g) = E(\int_T fg\, dt)$. It is enough to show that

$$(f,g) = (I(f), I(g)).$$

Let f,g be expressed by the division points (t_i). Then

$$(I(f), I(g)) = (\sum f_i X_i , \sum g_j X_j)$$

with $f_i = f(t_i)$, $g_j = g(t_j)$, $X_i = \beta(t_{i+1}) - \beta(t_i)$. Note that $f_i \in (\mathcal{B}^{t_i})$ and X_i is independent of $\mathcal{B}^{t_i}$. We have

$$(I(f), I(g)) = \sum_i E(f_i g_i)E(X_i^2) + \sum_{i<j} E(f_i g_i X_i)E(X_j) = \sum_i E(f_i g_i)(t_{i+1} - t_i)$$

$$= E\left[\sum f_i g_i (t_{i+1} - t_i)\right] = E(\int_T fg\, dt) = (f,g).$$

<u>Proof of (B)</u>. Let f be expressed by the division points (t_i) and put $f_i = f(t_i)$, $X_i = \beta(t_{i+1}) - \beta(t_i)$, $\Delta_i = t_{i+1} - t_i$ and $\delta = \| f \|_{\mathcal{L}_s}$. Then

$$P\left(\int_T f^2 dt > \epsilon^2\right) \leq \delta \frac{1+\epsilon}{\epsilon}.$$

Let

$$Y_i = \begin{cases} 1 & \text{if } \sum_{j=0}^{i} f_j^2 \Delta_j \leq \epsilon \\ 0 & \text{if } \sum_{j=0}^{i} f_j^2 \Delta_j > \epsilon . \end{cases}$$

Then $Y_i \in (\mathcal{B}^{t_i})$ and since X_i is independent of $\mathcal{B}^{t_i}$

$$E\left[\left(\sum_{i=0}^{n-1} Y_i f_i X_i\right)^2\right] = \sum_{i=0}^{n-1} E(Y_i^2 f_i^2)\Delta_i = E\left[\sum_{i=0}^{n-1} Y_i f_i^2 \Delta_i\right]$$

since from the definition of Y_i, $Y_i^2 = Y_i$. Again from the definition of Y_i, $\sum_{i=0}^{n-1} Y_i f_i^2 \Delta_i \leq \epsilon^2$ so that $E(S^2) \leq \epsilon^2$, where $S = \sum_{i=0}^{n-1} Y_i f_i X_i$. Now $P(|S| > \eta) \leq \epsilon^2/\eta^2$. If $\int_T f^2 dt \equiv \sum_i f_i^2 \Delta_i \leq \epsilon^2$ then $Y_0 = Y_1 = \ldots = Y_{n-1} = 1$ so that $S = \sum f_i X_i = I(f)$. Therefore

$$P(I(f) \neq S) \leq P\left(\int_T f^2\, dt > \epsilon^2\right) \leq \delta \frac{1+\epsilon}{\epsilon}$$

$$P(|I(f)| > \eta) \leq \delta \frac{1+\epsilon}{\epsilon} + \frac{\epsilon^2}{\eta^2}$$

and

$$\| I(f) \|_s = \int \frac{1}{1+|I(f)|} |I(f)|\, dP = \int_{|I(f)| \leq \eta} \frac{1}{1+|I(f)|} |I(f)|\, dP + \int_{|I(f)| > \eta} \frac{1}{1+|I(f)|} |I(f)|\, dP$$

$$\leq \eta + \delta\frac{1+\epsilon}{\epsilon} + \frac{\epsilon^2}{\eta^2}.$$

Putting $\epsilon = \delta^{2/3}$, $\eta = \epsilon^{\frac{1}{2}}$, we get $\|I(f)\|_s \leq 4\,\delta^{1/3}$.

Using linearity of I and the fact $\|I(f)\| = \|f\|$ for $f \in \mathcal{E}$, we can extend I to $\mathcal{L}^2(\|\ \|)$ [since $\mathcal{E}$ is dense in $\mathcal{L}^2(\|\ \|)$] such that I is linear. For $f \in \mathcal{L}^2$, $I(f) \in L^2(\Omega)$ and $\|f\| = \|I(f)\|$, and $E(I(f)) = 0$.

The linearity of I and the fact $\|I(f)\|_s \leq 4\|f\|_{\mathcal{L}_s}^{1/3}$ imply that we can extend I to the closure of $\mathcal{E}$ in $\|\ \|_{\mathcal{L}_s}$ i.e. to $\mathcal{L}_s$. Since for $f \in \mathcal{L}^2$, $\|f\|_{\mathcal{L}_s} \leq \|f\|$ we see that this extension coincides with the above for $f \in \mathcal{L}^2$. Further for $f \in \mathcal{L}_s$ we have $\|I(f)\|_s \leq 4\|f\|_{\mathcal{L}_s}^{1/3}$.

Using the remark at the end of the previous article we can show that, for $f \in \mathcal{L}^2$,

$$I(f) = \underset{n\to\infty}{\text{l.i.m}} \sum_i f^{M_n}(t_i^{(n)})\left[\beta(t_{i+1}^{(n)}) - \beta(t_i^{(n)})\right]$$

for some $\Delta_n = (t_i^{(n)})$ and M_n.

Finally if $f, g \in \mathcal{L}_s$ and if $f = g$ on a measurable set Ω_1, then $I(f) = I(g)$ a.e. on Ω_1.

§4. Definition of stochastic integral (3) - Continuous version.

Let $\mathcal{B}^t$, $0 \leq t < \infty$ be a monotone increasing system of Borel subalgebras of $\mathcal{B}$ such that $\mathcal{B}^t$ includes all null sets for each t. Let β_t, $0 \leq t < \infty$ be a Wiener process such that $\beta_t \in (\mathcal{B}^t)$ and $\beta_{t+h} - \beta_t$ is independent of $\mathcal{B}^t$ for $h > 0$.

Let $f_t = f_t(w) = f(t,w)$, $0 \leq t < \infty$ be such that

1. f is measurable in the pair (t,w).

2. $f_t \in (\mathcal{B}^t)$ for almost all t.

3. $\int_u^v f_t^2 \, dt < \infty$ for almost all $w \in \Omega$ for any finite interval $[u,v] \subset [o,\infty)$. Consider also the following conditions besides 1 and 2.

3'. $\int_\Omega \int_u^v f_t^2 \, dt \, dP < \infty$ for any finite interval $[u,v] \subset [o,\infty)$.

3''. There exist point $0 \leq t_0 < t_1 < t_2 < \dots \longrightarrow \infty$ and constants M_i independent of w such that

$$f_t(w) = f(t_i,w) \ , |f(t_i)| \leq M_i \ , \ t_i \leq t < t_{i+1} \ , \ i \geq 0.$$

In the same way as in § 2 we introduce **three** function classes $\mathcal{E}$, $\mathcal{L}^2$ and $\mathcal{L}$ as follows

$$\mathcal{E} = \{ f : 1,2,3'' \text{ hold} \}$$
$$\mathcal{L}^2 = \{ f : 1,2,3' \text{ hold} \}$$
$$\mathcal{L} = \{ f : 1,2,3 \text{ hold} \} .$$

From § 3 we can define $I(u,v) = \int_u^v f(t,w) d\beta(t,w)$, for $f \in \mathcal{L}$ and for any bounded interval $[u,v] \subset [o,\infty)$.

Now we shall show

<u>Theorem 1</u>. $I(u,v)$ has a continuous version in $[u,v]$ i.e. there exists $I(u,v)$ such that

$$P \left[I(u,v) = \int_u^v f d\beta \right] = 1 \quad \text{for any pair } (u,v)$$

and $I(u,v)$ is continuous in the pair (u,v) for almost all w ;

$I(u,v)$ is uniquely determined in the sense that if $I_i(u,v)$ $i=1,2$ satisfy the above conditions, then

$$P\left[I_1(u,v) = I_2(u,v) \text{ for all } u,v\right] = 1.$$

<u>Proof</u>. It is enough to show that $I(t,f) = \int_0^t f d\beta$ has a continuous (in t) version $I^*(t,f)$ in $0 \leq t \leq v$ for any given $v > 0$, because $I(u,v) = I(o,v) - I(o,u)$. If $f \in \mathcal{E}$ then $I(t)$ itself is such a version and

$$P\left(\sup_{o \leq t \leq v} |I(t,f)| > \epsilon\right) \leq \frac{1}{\epsilon^2} \|f\|^2 \tag{1}$$

where $\|f\|^2 = \int_o^v \int_\Omega f^2 \, dt \, dP$.

To prove (1) let the restriction of f to $[o,v)$ be expressed by the division set $\Delta = (o = t_o < t_1 < \ldots < t_n = v)$ and $s_o, s_1, \ldots$ be a dense set in $[o,v)$ such that $t_i = s_i$, $o \leq i \leq n$. Let now $\tau_1, \ldots, \tau_m (m \geqslant n)$ be a rearrangement of $s_o, s_1, \ldots, s_m$ in order of magnitudes. Then

$$I(\tau_i, f) = \sum_{j<i} f(\tau_j)(\beta(\tau_{j+1}) - \beta(\tau_j))$$

Using arguments similar to those employed in the proof of Kolmogoroff's inequality we can prove the following

<u>Lemma</u>. If $x_1, \ldots, x_n$, $y_1, \ldots, y_n$ are random variables satisfying

1) y_i is independent of $(x_1, \ldots, x_i, y_1, \ldots, y_{i-1})$

2) $E(y_i) = 0$ and $E(x_i^2)$, $E(y_i^2) < \infty$

then

$$P\left(\max_{1\leq k\leq n}\left|\sum_{i=1}^{k} x_i y_i\right| \geq \epsilon\right) \leq \frac{1}{\epsilon^2}\sum E(x_i^2)E(y_i^2).$$

Thus we have

$$P\left(\max_{0\leq i\leq n}|I(\tau_i,f)| > \epsilon\right) \leq \frac{1}{\epsilon^2}\int_\Omega\int_0^v f^2\,dt\,dP$$

i.e. $$P\left(\max_{0\leq i\leq n}|I(s_i,f)| > \epsilon\right) \leq \frac{1}{\epsilon^2}\|f\|^2$$

Letting $n \longrightarrow \infty$ we have (1).

Let C_s denote the space of all functions $(h(t,w), 0\leq t\leq v, w\in\Omega)$ which are continuous in $[o,v]$ and introduce the norm $\|\ \|_{C_s}$ by

$$\|h\|_{C_s} = E\left(\frac{1}{1+\sup\limits_{0\leq t\leq v}|h(t,w)|}\sup_{0\leq t\leq v}|h(t,w)|\right)$$

We shall prove that for $f\in\mathcal{E}$

$$\|I(f)\|_{C_s} = O(\|f\|_{\mathcal{L}_s}^{1/3}) \qquad (2)$$

where

$$\|f\|_{\mathcal{L}_s} = E\left[\frac{1}{1+\sqrt{\int_0^v |f|^2 dt}}\sqrt{\int_0^v f^2\,dt}\right]$$

Define $Y_i = 1$ if $\sum_{j=0}^{i-1} f_j^2\,(t_{j+1}-t_j) \leq \epsilon^2$ and $Y_i = 0$ if $\sum_{j=0}^{i-1} f_j^2(t_{j+1}-t_j) > \epsilon^2$ and let $g(t) = Y_i\,f(t) = Y_i f(t_i)$ for $t_i \leq t < t_{i+1}$. Then $g(t)\in(\mathcal{B}^t)$ and $\|g\|^2 = \int_\Omega\int_0^t g^2\,dt\,dP \leq \epsilon^2$

and

$$P(I(t,f)\neq I(t,g) \text{ for some } t\in[0,v)) = P\left(\int_0^v f^2dt > \epsilon^2\right) < \delta\,\frac{1+\epsilon}{\epsilon}$$

where $\delta = \|f\|_{\mathcal{L}_s}$. Thus

$$P\left(\sup_{0\leq t\leq v}|I(t,f)|>\eta\right)\leq P\left(\sup_{0\leq t\leq v}|I(t,g)|>\eta\right) + \delta\frac{1+\epsilon}{\epsilon}\leq \frac{\epsilon^2}{\eta^2} + \delta\frac{1+\epsilon}{\epsilon}$$

from (1). Therefore

$$\|I(.,f)\|_{C_s}\leq \eta + \delta\frac{1+\epsilon}{\epsilon} + \frac{\epsilon^2}{\eta^2}.$$

Putting $\epsilon = \delta^{2/3}$ and $\eta = \epsilon^{\frac{1}{2}}$, we get

$$\|I(.,f)\|_{C_s}\leq \epsilon^{\frac{1}{2}}[2+\epsilon+\epsilon^{\frac{1}{2}}]\leq 4\epsilon^{\frac{1}{2}} = O(\epsilon^{\frac{1}{2}}) = O(\delta^{1/3}) = O(\|f\|_{\mathcal{L}_s}^{1/3}).$$

Since C_s is complete in the norm $\|\ \|_{C_s}$ we can extend the mapping

$$\mathcal{E}\ni f\longrightarrow I(.,f)\in C_s$$

to the closure of $\mathcal{E}$ with respect to $\|\ \|_{\mathcal{L}_s}$ i.e. to $\mathcal{L}_s$. This extension gives the continuous version of $I(t,f)$, $0\leq t\leq v$. Since (2) is also true for this extension, we have

Theorem 2. If

$$\int_0^v |f_n - f|^2\, dt \longrightarrow 0 \text{ in probability}$$

then $\sup_{0\leq t\leq v}|I(t,f_n) - I(t,f)|\longrightarrow 0$ in probability.

For any Borel set $E\in[u,v)$ we define

$$\int_E f_\theta\, d\beta_\theta = \int_u^v f\chi_E\, d\beta_\theta .$$

For $f \in \mathcal{L}^2$ we have seen that

$$\| I(t,f) \| = \| f \| .$$

Let $f \in \mathcal{L}_s$ and consider the truncation f^M. Since $\int_u^v | f^M - f|^2 \chi_E \, ds \longrightarrow 0$, we see that

$$\sup_{u \leq t \leq v} | I(t, f\chi_E) - I(t, f^M \chi_E)| \longrightarrow 0 \text{ in probability.}$$

Since $\chi_E f^M \in \mathcal{L}^2$ we have, if E has Lebesgue measure zero,

$$\| I(t, f^M \chi_E) \| = \int_0^t E(f^M) \chi_E \, ds = 0.$$

Thus $\int_E f_\omega \, d\beta_\omega = 0$ if the Lebesgue measure of E is zero.

<u>Remark</u>. Henceforth when we speak of the stochastic integral we shall always understand it to mean the continuous version.

If $f \in \mathcal{L}^2$, we have

<u>Theorem 3</u>. If $f \in \mathcal{L}^2$ and $\epsilon > 0$,

$$P \left[||| I(t,f) ||| > \epsilon \right] \leq \frac{1}{\epsilon^2} \| f \|^2,$$

where $||| I(t,f) ||| = \sup_{0 \leq t \leq v} \| I(t,f) \|$.

<u>Proof</u>. Let $f_n \in \mathcal{E}$ be such that $\| f_n - f \| \longrightarrow 0$. Then for any $\delta > 0$,

$$P \left[||| I(t, f_n - f) ||| > \delta \right] \longrightarrow 0 \tag{1}$$

For $g \in \mathcal{E}$, we have proved that

$$P \left[||| I(t,g) ||| > \epsilon \right] \leq \frac{1}{\epsilon^2} \| g \|^2 \tag{2}$$

Therefore if $\eta > 0$,

$$P[|||I(t,f)||| > \epsilon + \eta] \leq P[|||I(t,f_n - f)||| + |||I(t,f_n)||| > \epsilon + \eta]$$

$$\leq P[|||I(t,f_n - f)|||] + P[|||I(t,f_n)||| > \epsilon] \leq P[|||I(t,f_n - f)||| > \eta] + \\ + \frac{1}{\epsilon^2} \|f_n\|^2$$

from (2). From (1) if $n \longrightarrow \infty$, $P[|||I(t,f)||| > \epsilon + \eta]$
$\leq \frac{1}{\epsilon^2} \|f\|^2$. Letting $\eta \longrightarrow 0$, we get the result.

§5. Stochastic differentials.

Let $\mathcal{B}^t, \beta_t$ be defined as before. If $x_t = x_o + \int_o^t f_s \, d\beta_s + \int_o^t g_s \, ds$, where $x_o(w) \in \mathcal{B}^o$ and

1. f,g are measurable in the pair (t,w)
2. $f_s, g_s \in (\mathcal{B}^s)$ for almost all $s, 0 \leq s < \infty$
3. $\int_o^t f_s^2 \, ds < \infty$, $\int_o^t |g_s| \, ds < \infty$ for almost all w, for any finite t,

then we write

$$dx_t = f_t \, d\beta_t + g_t \, dt.$$

If $dx_t = f_t \, d\beta_t + g_t \, dt$, $dx_t^i = f_t^i \, d\beta_t + g_t^i \, dt$, $f_t = \sum_i \varphi_t^i f_t^i$, $g_t = \sum_i \varphi_t^i g_t^i + \psi(t)$ then we shall write

$$dx_t = \sum_i \varphi_t^i \, dx_t^i + \psi_t \, dt.$$

<u>Theorem</u>. If $F(\xi^1, \ldots, \xi^k, t)$ is C^2 in $(\xi^1, \ldots, \xi^k)$ and C^1 in t, if $dx_t^i = f_t^i \, d\beta_t + g_t^i dt$ and if $y_t = F(x_t^1, \ldots, x_t^k, t)$, then

$$dy_t = \sum_i F_i dx_t^i + \Big[\tfrac{1}{2} \sum_{i,j=1}^k F_{ij} \, f_t^i f_t^j + F_{k+1}\Big] dt$$

where $F_i = \frac{\partial F}{\partial \xi_i}$, $F_{ij} = \frac{\partial^2 F}{\partial \xi^i \partial \xi^j}$, $F_{k+1} = \frac{\partial F}{\partial t}$.

Remark. We can get the result formally as follows:

1. Expand dy_t i.e. $dy_t = dF(x_t^1,\dots,x_t^k,t) = \sum_i F_i\, dx_t^i + F_{k+1}dt + \frac{1}{2}\sum_{i,j=1}^{k} f_{ij}\, dx_t^i\, dx_t^j + \dots$
2. Put $dx_t^i = f_t^i\, d\beta_t + g_t^i\, dt$.
3. Use $d\beta_t \approx \sqrt{dt}$
4. Ignore $o(dt)$.

Lemma 1. If $f,g \in \mathcal{L}_s$ (as defined in § 2) then

$$\left[\int_t^u f_s\, d\beta_s\right]\left[\int_t^u g_s\, d\beta_s\right] = \int_t^u f_s G_s d\beta_s + \int_t^u g_s F_s d\beta_s + \int_t^u f_s g_s ds,$$

where

$$F_s = \int_t^s f_\theta\, d\beta_\theta\,,\; G_s = \int_t^s g_\theta\, d\beta_\theta\,.$$

Proof. Case 1. $f,g \in \mathcal{E}$ (as defined in § 2).

We can express f and g by the same set of division points $\Delta = (t = t_0 < t_1 \dots < t_m = u)$. Now let $\Delta_n = (t = t_0^{(n)} < t_1^{(n)} < \dots < t_n^{(n)} = u)(n \geq m)$ be a sequence of sets of division points containing Δ such that $\delta(\Delta_n) = \max_{0 \leq i \leq n-1} |t_{i+1}^{(n)} - t_i^{(n)}| \longrightarrow 0$. Put $X_i^{(n)} = f(t_i^{(n)})$, $Y_i^{(n)} = g(t_i^{(n)})$, $B_i^{(n)} = \beta(t_{i+1}^{(n)}) - \beta(t_i^{(n)})$. We have

$$\left[\int_t^u f_s d\beta_s\right]\left[\int_t^u g_s d\beta_s\right] = \left[\sum_{i=0}^{n-1} X_i^{(n)} B_i^{(n)}\right]\left[\sum_{j=0}^{n-1} Y_j^{(n)}\, B_j^{(n)}\right]$$

$$= \sum_{i=1}^{n-1} X_i^{(n)} G(t_i^{(n)}) B_i^{(n)} + \sum_{i=1}^{n-1} Y_i^{(n)} F(t_i^{(n)}) B_i^{(n)} + \sum_{i=0}^{n-1} X_i^{(n)} Y_i^{(n)} (B_i^{(n)})^2.$$

Put $\varphi_n(s) = t_i^{(n)}$ for $t_i^{(n)} \leq s < t_{i+1}^{(n)}$ and let G_n, F_n be defined as

$$G_n(s,w) = G(\varphi_n(s),w) , F_n(s,w) = F(\varphi_n(s),w).$$

Then G_n, $F_n \in \mathcal{E}$ and since the set Δ_n contains Δ, $f G_n$, $g F_n \in \mathcal{E}$. Thus

$$\int_t^u f_s d\beta_s \int_t^u g_s d\beta_s = \int_t^u f(s)G_n(s)d\beta_s + \int_t^u g_s F_n(s)d\beta_s + \sum_{i=0}^{n-1} X_i^{(n)} Y_i^{(n)} (B_i^{(n)})^2.$$

Now $\int_t^u |f(s)G(\varphi_n(s))-f(s)G(s)|^2 ds \leq \max_{t \leq s \leq u} |G_n(s)-G(s)| \int_t^u |f(s)|^2 ds \to 0$

with probability 1 since $G(s)$ is continuous in s. Similarly

$$\int_t^u |g(s)F_n(s)-g(s)F(s)|^2 ds \to 0 \text{ with probability } 1.$$

Eurther

$$E\left[\left(\sum_{i=0}^{n-1} X_i^{(n)} Y_i^{(n)} \left[(B_i^{(n)})^2 - (t_{i+1}^{(n)} - t_i^{(n)})\right]\right)^2\right]$$

$$= \sum_{i=0}^{n-1} E((X_i^{(n)})^2 (Y_i^{(n)})^2) E\left[\left(\left[(B_i^{(n)})^2 - (t_{i+1}^{(n)} - t_i^{(n)})\right]\right)^2\right] +$$

$$+2 \sum_{i<j} E\left\{X_i^{(n)} Y_i^{(n)} X_j^{(n)} Y_j^{(n)} \left[(B_i^{(n)})^2 - (t_{i+1}^{(n)} - t_i^{(n)})\right] \left[(B_j^{(n)})^2 - (t_{j+1}^{(n)} - t_j^{(n)})\right]\right\}$$

$$= \sum_{i=0}^{n-1} E((X_i^{(n)} Y_i^{(n)})^2) E\left[\left((B_i^{(n)})^2 - (t_{i+1} - t_i^{(n)})\right)^2\right] , \text{ since}$$

$$E((B_j^{(n)})^2) = t_{j+1}^{(n)} - t_j^{(n)}$$

$$=2\sum_{i=0}^{n-1} E((X_i^{(n)}Y_i^{(n)})^2)(t_{i+1}^{(n)}-t_i^{(n)})^2 \leq 2\delta(\Delta_n)E\Big[\sum_{i=0}^{n-1}(X_i^{(n)}Y_i^{(n)})^2(t_{i+1}^{(n)}-t_i^{(n)})\Big]$$

$$= 2\delta(\Delta_n)E(\int_t^u f^2(s)g^2(s)ds), \text{ since } \sum_{i=0}^{n-1}(X_i^{(n)}Y_i^{(n)})^2(t_{i+1}^{(n)}-t_i^{(n)}) =$$

$$= \int_t^u f^2(s)g^2(s)ds.$$

The lemma for $f,g \in \mathcal{E}$, then follows from Theorem 2 of § 4.

Case 2. Let $f,g \in \mathcal{L}_s$. There exist sequences $f_n, g_n \in \mathcal{E}$ such that

$$\int_t^u |f_n - f|^2 ds \text{ and } \int_t^u |g_n - g|^2 ds \longrightarrow 0 \text{ in probability.}$$

Therefore

$$\sup_{t \leq s \leq u} |F_n - F| \text{ and } \sup_{t \leq s \leq u} |G_n - G| \longrightarrow 0 \text{ in probability}$$

where $F(s) = \int_t^s f(\theta)d\beta_\theta$, $G(s) = \int_t^s g(\theta)d\beta_\theta$, $F_n(s) =$ $= \int_t^s f_n(\theta)d\beta_\theta$, $g_n(s) = \int_t^s g_n(\theta)d\beta_\theta$. Choosing a subsequence if necessary we can assume that the above limits are true almost everywhere. Then for any w

$$\int_t^u |f_nG_n - fG|^2 ds \leq 2\int_t^u |f_n - f|^2 G_n^2 ds + 2\int_t^u f^2 |G_n - G|^2 ds$$

$$\leq 2 \sup_{t \leq s \leq u} G_n^2(s)\int_t^u |f_n - f|^2 ds + 2 \sup_{t \leq s \leq u} |G_n - G|^2 \int_t^u f^2 ds \longrightarrow 0.$$

The proof of the lemma can be completed easily.

Proceeding on the same lines and noting that

$$\sum_i f(t_i^{(n)})g(t_i^{(n)})(\beta(t_{i+1}^{(n)}) - \beta(t_i^{(n)}))(t_{i+1}^{(n)} - t_i^{(n)}) \longrightarrow 0, \text{ for } f,g \in \mathcal{E},$$

as $n \longrightarrow \infty$ we can prove

<u>Lemma 2</u>. $$\left(\int_t^u f_s d\beta_s\right)\left(\int_t^u g_s ds\right) = \int_t^u f_s G_s d\beta_s + \int_t^u g_s F_s ds$$

where $F_s = \int_t^s f_\theta\, d\beta_\theta$, $G_s = \int_t^s g_\theta\, d\theta$.

<u>Proof of the theorem</u>. Write $F(x_t^1,\dots,x_t^k,t) = F(x_t)$. Let $\Delta^n = (0=t_0^{(n)} < t_1^{(n)} < t_1^{(n)} < \dots < t_n^{(n)} = t)$ be a sequence of subdivisions such that $\delta(\Delta_n) \longrightarrow 0$. Then

$$\begin{aligned} y_t = y_0 &+ \sum_{l=0}^{n-1}\sum_{i=1}^{k} F_i(x(t_l^{(n)}))(x^i(t_{l+1}^{(n)})-x^i(t_l^{(n)})) + \sum_{l=0}^{n-1} F_{k+1}(x(t_l^{(n)}))(t_{l+1}^{(n)}-t_l^{(n)}) \\ &+ \tfrac{1}{2}\sum_{l=0}^{n-1}\sum_{i,j=1}^{k} F_{ij}(x(t_l^{(n)}))(x^i(t_{l+1}^{(n)})-x^i(t_l^{(n)}))(x^j(t_{l+1}^{(n)})-x^j(t_l^{(n)})) \\ &+ \tfrac{1}{2}\sum_{l=0}^{n-1}\sum_{i,j=1}^{k} \epsilon_{ijl}^{(n)}(x^i(t_{l+1}^{(n)})-x^i(t_l^{(n)}))(x^j(t_{l+1}^{(n)})-x^j(t_l^{(n)})) \\ = y_0 &+ \sum_{i=1}^{k} I_{in}^1 + I_n^2 + \tfrac{1}{2}\sum_{i,j=1}^{k} I_{ijn}^3 + \tfrac{1}{2}\sum_{i,j=1}^{k} I_{ijn}^4 \text{, say.}\end{aligned}$$

From the hypotheses on F and the continuity of $x^j(t)$,

$$\epsilon_{ijl}^{(n)} \longrightarrow 0 \text{ uniformly in } i,j,l \text{ as } n \longrightarrow \infty .$$

Let $\varphi_n(t) = t_l^{(n)}$ for $t_l^{(n)} \leq t < t_{l+1}^{(n)}$. Then we have

$$\begin{aligned} I_{in}^1 &= \sum_{l=0}^{n-1} F_i(x(t_l^{(n)}))\left[\int_{t_l^{(n)}}^{t_{l+1}^{(n)}} f_s^i\, d\beta_s + \int_{t_l^{(n)}}^{t_{l+1}^{(n)}} g_s^i\, ds\right] \\ &= \sum_{l=0}^{n-1}\left[\int_{t_l^{(n)}}^{t_{l+1}^{(n)}} F_i(x(\varphi_n(s))) f_s^i d\beta_s + \int_{t_l^{(n)}}^{t_{l+1}^{(n)}} F_i(x(\varphi_n(s))) g_s^i\, ds\right]\end{aligned}$$

$$= \int_0^t F_i(x(\varphi_n(s)))f_s^i d\beta_s + \int_0^t F_i(x(\varphi_n(s)))g_s^i ds.$$

Also

$$\int_0^t |F_i(x(\varphi_n(s)))-F_i(x(s))|^2 (f_s^i)^2 ds \leq \max_{0\leq s\leq t} |F_i(x(\varphi_n(s)))-F_i(x(s))|^2 \times$$

$$\times \int_0^t (f_s^i)^2 ds \longrightarrow 0$$

for every w. Thus

$$\sum_{i=1}^k I_{in}^1 \longrightarrow \sum_{i=1}^k \left[\int_0^t F_i(x(s))f_s^i d\beta_s + \int_0^t F_i(x(s))g_s^i ds \right]$$

in probability. Similarly

$$I_n^2 = \int_0^t F_{k+1}(x(\varphi_n(s)))ds \longrightarrow \int_0^t F_{k+1}(x(s))ds.$$

Using Lemma 1 and 2, we have

$$(x^i(v)-x^i(u))(x^j(v)-x^j(u)) = \int_u^v \left[f_s^i(x^j(s)-x^j(u))+f_s^j(x^i(s)-x^i(u)\right] d\beta_s$$

$$+\int_u^v f_s^i f_s^j ds+\int_u^v g_s^i(\int_u^s f_\theta^j d\theta)ds+\int_u^v g_s^j(\int_u^s f_\theta^i d\theta)+(\int_u^v g_s^i ds)(\int_u^v g_s^j ds)$$

$$= \int_u^v \left[f_s^i(x^j(s)-x^j(u))+f_s^j(x^i(s)-x^i(u))\right] d\beta_s + \int_u^v f_s^i f_s^j ds$$

$$+ \int_u^v \left[g_s^i(Y_s^j-Y_u^j)+g_s^j(Y_s^i-Y_u^i)\right]ds$$

$$\left[\text{since } (\int_u^v g_s^i ds)(\int_u^v g_s^j ds) = \int_u^v g_s^i(\int_u^s g_\theta^j d\theta)ds+\int_u^v g_s^j(\int_u^s g_\theta^i d\theta)ds\right]$$

where $Y^i_s = \int_0^s \left[f^i_\theta + g^i_\theta\right] d\theta$, $Y^j_s = \int_0^s \left[f^j_\theta + g^j_\theta\right]\, d\theta$.

Thus

$$I^3_{ijn} = \int_0^t F_{ij}(x(\varphi_n(s)))\left[f^i_s(x^j(s)-x^j(\varphi_n(s)))+f^j_s(x^i(s)-x^i(\varphi_n(s)))\right] d\beta_s$$

$$+\int_0^t F_{ij}(x(\varphi_n(s)))f^i_s f^j_s ds+\int_0^t F_{ij}(x(\varphi_n(s)))\left[g^i_s(Y^j(s)-Y^j(\varphi_n(s))) + \right.$$

$$\left. + g^j_s(Y^i(s)-Y^i(\varphi_n(s)))\right] ds$$

$$\longrightarrow \int_0^t F_{ij}(x(s))\, f^i_s\, f^j_s\, ds \quad \text{in probability}$$

because other terms can, without difficulty, be shown, to tend to zero in probability. Again

$$|I^4_{ijn}| \leq \max_{0 \leq l \leq n-1} |\epsilon^{(n)}_{ijl}| \sum_{l=0}^{n-1} |x^i(t^{(n)}_{l+1})-x^i(t^{(n)}_l)||x^j(t^{(n)}_{l+1})-x^j(t^{(n)}_l)|$$

$$\leq \tfrac{1}{2} \max_{0 \leq l \leq n-1} |\epsilon^{(n)}_{ijl}| \sum_{l=0}^{n-1}\left[(x^i(t^{(n)}_{l+1})-x^i(t^{(n)}_l))^2+(x^j(t^{(n)}_{l+1})-x^j(t^{(n)}_l))^2\right]$$

In the same way as above, we can show that

$$\sum_{l=0}^{n-1} (x^i(t^{(n)}_{l+1})-x^i(t^{(n)}_l))^2 \longrightarrow \int_0^t f^i_s\, f^i_s\, ds.$$

Thus $|I^4_{ijn}| \longrightarrow 0$ in probability. We have proved the theorem

§6. Stochastic differential equations.

The notation in this article is as in the previous ones.

Theorem 1. Let $p(\xi)$, $r(\xi)$, $\xi \in R^1$ satisfying Lipschitz condition

$$|p(\xi)-p(\eta)| \leq A|\xi-\eta|, |r(\xi)-r(\eta)| \leq A|\xi-\eta|.$$

Then

$$dx_t = p(x_t)d\beta_t + r(x_t)dt, \quad x_o(w) = \alpha(w) \in (\mathcal{B}^o)$$

has one and only one solution.

$$\left[|\alpha(w)| < \infty \quad \text{for almost all} \quad w\right]$$

Proof: (a) Existence. We show that

$$x_t(w) = \alpha(w) + \int_o^t p(x_s)d\beta_s + \int_o^t r(x_s)ds$$

has a solution. We use successive approximation to get a solution. Let $\alpha^M(w)$ be the truncation of α at M (i.e. $(\alpha \vee (-M)) \wedge M$) and put

$$x^o(t,w) \equiv \alpha^M(w).$$

Define by induction on k

$$x^{k+1}(t,w) = \alpha^M(w) + \int_o^t p(x_s^k)d\beta_s + \int_o^t r(x_s^k)ds = \alpha^M + y^k(t) + z^k(t), \text{ say.}$$

Note that if $f \in \mathcal{L}^2$ then $I(t,f) \in \mathcal{L}^2$ and $E(|I(t,f)|^2) = \int_o^t E(|f(s)|^2)ds$, where $I(t,f) = \int_o^t f_s \, d\beta_s$. From the hypotheses on p and r, $x^k(t,w) \in \mathcal{L}^2$ for all k. Now

$$E(|x^{k+1}(t)-x^k(t)|^2) \leq 2E\left[|y^k(t)-y^{k-1}(t)|^2\right] + 2E\left[|z^k(t)-z^{k-1}(t)|^2\right]$$

$$\leq 2\int_o^t E(|p(x^k(s))-p(x^{k-1}(s))|^2)ds + 2t\, E\left(\int_o^t |r(x^k(s))-r(x^{k-1}(s))|^2 ds\right)$$

$$\left[\text{since } |z^k(t)-z^{k-1}(t)|^2 \leq t \int_0^t |r(x^k(s))-r(x^{k-1}(s))|^2 ds\right]$$

$$\leq 2A^2(1+t)\int_0^t E(|x^k(s)-x^{k-1}(s)|^2)ds \leq 2A^2(1+v)\int_0^t E(|x_s^{k-1}-x_s^k|^2)ds$$

where $0 \leq t \leq v < \infty$ and v is fixed for the present. Therefore

$$E(|x^{k+1}(t)-x^k(t)|^2) \leq [2A^2(1+v)]^k \int_0^t ds_1 \int_0^{s_1} ds_2 \ldots \int_0^{s_{k-1}} E(|x^1(s_k)-x^0(s_k)|^2)ds_k$$

$$\leq [2A^2(1+v)]^k \int_0^t ds_1 \ldots \int_0^{s_{k-1}} 2E(p^2(\alpha^M)s_k + r^2(\alpha^M)s_k^2)ds_k$$

$$= [2A^2(1+v)]^k \, 2\left[E(p^2(\alpha^M))\frac{t^{k+1}}{(k+1)!} + 2E(r^2(\alpha^M))\frac{t^{k+2}}{(k+2)!}\right]$$

which gives

$$\int_0^v E(|x^{k+1}(\theta)-x^k(\theta)|^2)d\theta \leq 2[2A^2(1+v)]^k\left[E(p^2(\alpha^M))\frac{v^{k+2}}{(k+2)!} + \right.$$

$$\left. + 2E(r^2(\alpha^M))\frac{v^{k+3}}{(k+3)!}\right]$$

Let $|||F(t,w)||| = \sup_{0 \leq t \leq v} |F(t,w)|$. Then

$$P\left[|||x^{k+1}(t)-x^k(t)||| > \epsilon\right] \leq P\left[|||y^k(t)-y^{k-1}(t)||| > \frac{\epsilon}{2}\right] + P\left[|||z^k(t)-z^{k-1}(t)||| > \frac{\epsilon}{2}\right]$$

$$\leq \frac{4}{\epsilon^2}\int_0^v E\left[|p(x^k(s))-p(x^{k-1}(s))|^2\right]ds + \frac{4}{\epsilon^2} v \int_0^v E\left[|r(x^k(s))-r(x^{k-1}(s))|^2\right]ds,$$

(from Theorem 3 of § 4),

$$\leq \frac{4A^2(1+v)}{\epsilon^2} 2\,[2A^2(1+v)]^{k-1}\left[E(p^2(\alpha^M))\frac{v^{k+1}}{(k+1)!} + 2E(r^2(\alpha^M))\frac{v^{k+2}}{(k+2)!}\right]$$

$$< \frac{B}{\epsilon^2}\frac{[2A^2v(1+v)]^k}{k!} \quad \text{where } B = 2\left[E(p^2(\alpha^M))v + 2v^2E(r^2(\alpha^M))\right].$$

Putting $\epsilon_k = \frac{[2A^2v(1+v)]^{k/3}}{(k!)^{1/3}}$, we get

$$P\left[|||x^{k+1}(t)-x^k(t)||| > \epsilon_k\right] \leq B\,\epsilon_k.$$

Since $\sum \epsilon_k$ is a convergent series Borel-Cantelli lemma implies that, with probability 1, w belongs only to a finite number of sets in the bracket of the last inequality. Therefore

$$P\left[|||x^{k+1}(t)-x^k(t)||| < \epsilon_k \text{ for all } k \geqslant \text{some } l\right] = 1.$$

Since $\sum \epsilon_k < \infty$, we get

$$|||x^m(t)-x^n(t)||| \longrightarrow 0 \text{ with probability } 1 \text{ as } m,n \longrightarrow \infty$$

i.e. $\qquad P\left[x^k(t) \text{ converges uniformly for } 0 \leq t \leq v\right] = 1.$

Taking $v = 1,2,3,\ldots$ we get

$$P\left[(x^k(t) \text{ converges uniformly for } 0 \leq t \leq n) \text{ for every } n\right] = 1.$$

Let $x^M(t,w)$ be the limit of $x^k(t,w)$. This is clearly continuous in t for almost all w. Also for any $v < \infty$,

$$P\left[|||x^k(t)-x^M(t)||| \longrightarrow 0 \text{ as } k \longrightarrow \infty\right] = 1$$

so that $\int_0^t p(x^k(s))d\beta_s \longrightarrow \int_0^t p(x^M(s))d\beta_s$ in probability.

Now we prove without difficulty that

$$x^M(t,w) = \alpha^M(w) + \int_0^t p(x^M(s,w))d\beta(s,w) + \int_0^t r(x^M(s,w))ds.$$

Let $\Omega_M = (w : |\alpha(w)| \leq M)$ and define $x(t,w) = x^M(t,w)$ on Ω_M. If $M < M'$, then on Ω_M, $\alpha^M = \alpha^{M'}$ so that from the construction [and the fact that if $f = g$ on a measurable set B then $I(t,f) = I(t,g)$ a.e. on B] it follows that $x^M(t,w) = x^{M'}(t,w)$. Also since on Ω_M, $x(t,w) = x^M(t,w)$, $x(t,w)$ is a solution.

(b) Uniqueness. Let

$$x_t = a + \int_0^t p(x_s)d\beta_s + \int_0^t r(x_s)ds$$

$$\qquad 0 \leq t \leq v,\ a \in (\mathcal{B}^0).$$

$$y_t = a + \int_0^t p(y_s)d\beta_s + \int_0^t r(y_s)ds$$

Case 1. $E(x_t^2)$ and $E(y_t^2)$ are bounded by some $G < \infty$ for $0 \leq t \leq v$. We have

$$E((x(t)-y(t))^2) \leq 2E(\left[\int_0^t (p(x(s))-p(y(s)))d\beta(s)\right]^2)$$

$$+ 2E(\left[\int_0^t (r(x(s))-r(y(s)))ds\right]^2)$$

$$\leq 2\int_0^t E\left[(p(x(s))-p(y(s)))^2\right]ds + 2t\int_0^t E\left[(r(x(s))-r(y(s)))^2\right]ds$$

since $\left[\int_0^t \varphi(s)ds\right]^2 \leq t\int_0^t \varphi^2(s)ds$. Thus

$$E\left[(x_t-y_t)^2\right] \leq 2A^2(1+t)\int_0^t E\left[(x_s-y_s)^2\right]ds \leq 2A^2(1+v)\int_0^t E\left[(x_s-y_s)^2\right]ds$$

Put $C_t = E((x_t-y_t)^2)\ [\leq 4G^2]$. Then

$$C_t \leq 2A^2(1+v)\int_0^t C_s\,ds \leq \left[2A^2(1+v)\right]^2 \int_0^t ds\int_0^s C_\theta\,d\theta \leq \cdots$$

Therefore

$$C_t \leq \frac{[2A^2(1+v)]^n}{n!} t^n \, 4\, G^2 \to 0 \quad \text{as} \quad n \to \infty .$$

Case 2. Let $x_{tM} = (x_t \wedge M) \vee (-M)$, $y_{tM} = (y_t \wedge M) \vee (-M)$ and $a_M = (a \vee (-M)) \wedge M$. Define $x^0, x^1, \ldots, y^0, y^1, \ldots$ inductively as follows :

$$x_t^0 = x_{tM}, x_t^{n+1} = a_M + \int_0^t p(x_s^n)\, d\beta_s + \int_0^t r(x_s^n)\, ds,$$

$$y_t^0 = y_{tM}, y_t^{n+1} = a_M + \int_0^t p(y_s^n)\, d\beta_s + \int_0^t r(y_s^n)\, ds.$$

Arguments similar to those used in the proof of existence of a solution prove that

$$\tilde{x}_t = \lim_{n\to\infty} x_t^n, \tilde{y}_t = \lim_{n\to\infty} y_t^n$$

exist and

$$\tilde{x}_t = a_M + \int_0^t p(\tilde{x}_s)\, d\beta_s + \int_0^t r(\tilde{x}_s)\, ds,$$

$$\tilde{y}_t = a_M + \int_0^t p(\tilde{y}_s)\, d\beta_s + \int_0^t r(\tilde{y}_s)\, ds$$

and

$$\sup_{0\leq t\leq v} E(\tilde{x}_t^2) < \infty, \quad \sup_{0\leq t\leq v} E(\tilde{y}_t^2) < \infty .$$

Therefore from Case 1, $\tilde{x}_t = \tilde{y}_t$ for $0 \leq t \leq v$.

$$\text{Let } \Omega_M = (w : |a| < M, \sup_{0\leq t\leq v} |x_t| < M, \sup_{0\leq t\leq v} |y_t| < M).$$

Then since x_t and y_t have continuous paths, $P[\bigcup_M \Omega_M] = 1$. But on Ω_M,

$$\left.\begin{array}{l} y_t = y_{tM} = y_t^o = y_t^1 = \dots \\ x_t = x_{tM} = x_t^o = x_t^1 = \dots \end{array}\right\} \quad 0 \leq t \leq v$$

Note that if $f,g \in \mathcal{L}_s$ and $f = g$ on a measurable subset B then

$$I(t,f) = I(t,g) \quad \text{on } B \text{ with probability } 1 \quad .$$

Thus $x_t = y_t$ on Ω_M. We have proved the theorem.

<u>Corollary</u>. Let $\alpha(w) \in L^2(\Omega)$ and $x(t,w)$ satisfy

$$x(t,w) = \alpha(w) + \int_0^t p(x(s,w))d\beta(s,w) + \int_0^t r(x(s,w))ds \ .$$

Then

$$E(x_t^2) \leq \beta e^{\mu t} \quad \text{for } 0 \leq t \leq v$$

where $\beta = 3E(\alpha^2) + 6v\, p^2(0) + 6v^2 r^2(0)$ and $\mu = 6A^2(1+v)$.

<u>Proof</u>. From the proof of Theorem 1 we gather that $x(t,w) \in \mathcal{L}^2$ (for any $v < \infty$). If $\| x(t) \|^2 = E(|x(t)|^2)$,

$$\| x(t)-x(s) \|^2 \leq 2 \int_s^t E(p(x(\theta))^2)d\theta + 2(t-s) \int_s^t E(r(x(\theta))^2)d\theta$$

so that $\| x(t) \|^2$ is continuous in t. Let $l = \sup_{0 \leq t \leq v} \| x_t \|^2$.

Now

$$\begin{aligned} \| x_t \|^2 &\leq 3E(\alpha^2)+3 \int_0^t E(p(x(s))^2)ds+3t \int_0^t E(r(x(s))^2)ds \\ &\leq 3E(\alpha^2)+6 \left[tp^2(o)+t^2r^2(o)\right]+6A^2(1+t) \int_0^t \| x(s_1) \|^2 ds_1 . \\ &\leq \beta + \mu \int_0^t \| x(s_1) \|^2 ds_1 \leq \beta + \mu \int_0^t ds_1 \left[\beta + \mu \int_0^{s_1} \| x(s_2) \|^2 ds_2 \right] \end{aligned}$$

$$= \beta(1+\mu t) + \mu^2 \int_0^t ds_1 \int_0^{s_1} \|x(s_2)\|^2 ds_2 \leq \beta(1+\mu t) + \mu^2 \int_0^t ds_1 \times$$

$$\times \int_0^{s_1} ds_2 \Big[\beta + \mu \int_0^{s_2} \|x(s_3)\|^2 ds_3\Big]$$

$$= \beta\Big[1 + \mu t + \frac{\mu^2 t^2}{2!}\Big] + \mu^3 \int_0^t ds_1 \int_0^{s_1} ds_2 \int_0^{s_2} \|x(s_3)\|^2 ds_3$$

Continuing this, for any n we have

$$\|x(t)\|^2 \leq \beta\Big[1 + \mu t + \frac{\mu^2 t^2}{2!} + \ldots + \frac{\mu^n t^n}{n!}\Big] + \mu^{n+1} \int_0^t ds_1 \int_0^{s_1} ds_2 \ldots$$

$$\ldots \int_0^{s_n} \|x(s_{n+1})\|^2 ds_{n+1}$$

$$\leq \beta e^{\mu t} + \mu^{n+1} \int_0^t ds_1 \ldots \int_0^{s_n} \|x(s_{n+1})\|^2 ds_{n+1} \leq \beta e^{\mu t} + \mu^{n+1} I \frac{t^{n+1}}{(n+1)!}$$

Q.E.D.

Theorem 2. There exists a function $x(t,a,w)$ measurable in the pair (a,w) such that for every fixed $a \in R^1$,

$$x(t,a,w) = a + \int_0^t p(x(s,a,w))d\beta(s,w) + \int_0^t r(x(s,a,w))ds$$

for every t and for almost all w. That is there exists a version of the solution of

$$dx_t = p(x_t)d\beta_t + r(x_t)dt, \; x(o) = a$$

which is measurable in the pair (a,w).

Proof. Let $x^o(t,a,w) \equiv a$. $x^o(t,a,w)$ is measurable in the pair (a,w).

Assume that $x^1,\ldots,x^k$ have been defined, are measurable in the pair (a,w) and for every fixed a

$$x^i(t,a,k) = a+\int_0^t p(x^{i-1}(s,a,w))d\beta(s,w)+\int_0^t r(x^{i-1}(s,a,w))ds,\qquad 1\leq i\leq k.$$

for almost all w and for all t. We shall define x^{k+1}. Let $\Delta^n = (0 = s_0^n < s_1^n < \ldots)$ be a sequence of subdivisions of $[0,\infty)$ such that $\delta_n = \sup_i |s_{i+1}^n - s_i^n|$ tends to zero. Let $v<\infty$. Then since $x^k(s,a,w)$ is continuous in s for almost all w,

$$\int_0^v |p(x^k(s,a,w))-p(x^k(\varphi_n(s),a,w))|^2 ds \longrightarrow 0$$

for almost all w, where $\varphi_n(t) = s_i^n$ for $s_i^n \leq t < s_{i+1}^n$. Hence

$$\sup_{0\leq t\leq v}\left|\int_0^t \left[p(x^k(s,a,w))-p(x^k(\varphi_n(s),a,w))\right]d\beta_s\right| \longrightarrow 0$$

in probability. By the diagonal process, we can find a subsequence n_j such that

$$P\left[\sup_{0\leq t\leq v}\left|\int_0^t p(x^k(s,a,w))d\beta_s - \int_0^t p(x^k(\varphi_{n_j}(s),a,w))d\beta_s\right| \to 0 \text{ for every } v<\infty\right] = 1$$

Since $p(x^k(\varphi_{n_j}(s),a,w))\in \mathcal{E}$,

$$\int_0^t p(x^k(\varphi_{n_j}(s),a,w) \text{ is measurable in } (a,w).$$

It follows that $M(t,a,w) = \overline{\lim}\int_0^t p(x^k(\varphi_{n_j}(s),a,w))d\beta_s$ is measurable in (a,w). Now define

$$x^{k+1}(t,a,w) = a + M(t,a,w) + \int_0^t r(x^k(s,a,w))ds.$$

Proceeding as in Theorem 1 we can show that $x^k(t,a,w)$ converges with probability 1. Let now

$$x(t,a,w) = \overline{\lim}\, x^k\,(t,a,w)$$

We can show that $x(t,a,w)$ is the required function.

Remark. We can easily prove that if $a_n \longrightarrow a$ then $x(t,a_n,w) \longrightarrow x(t,a,w)$ in probability. In fact

$$E(|x(t,a,w)-x(t,a_n,w)|^2) \leq 3|a-a_n|^2+3\int_0^t E(|p(x(s,a_n,w)-p(x(s,a,w))|^2)ds$$
$$+\ 3t\int_0^t E(|r(x(s,a_n,w))-r(x(s,a,w))|^2)ds$$

so that

$$\overline{\lim_{a_n\to a}}\ E\left[|x(t,a,w)-x(t,a_n,w)|^2\right] \leq 3A^2(1+t)\ \overline{\lim_{a_n\to a}}\int_0^t E(|x(s,a_n,w)-x(s,a,w)|^2)ds$$

and now using the corollary to Theorem 1 and Fatou lemma we get

$$\overline{\lim_{a_n\to a}}\ E\left[|x(t,a,w)-x(t,a_n,w)|^2\right] \leq 3A^2(1+t)\int_0^t \overline{\lim_{a_n\to a}}\ E(|n(s,a_n,w)-x(s,a,w)|^2)ds$$

Q.E.D.

Theorem 3. Let $x(t,a,w)$ be as in Theorem 2 and $x(t,w)$ be the solution of

$$dx_t = p(x_t)d\beta_t + r(x_t)dt\ ,\ x(o,w) \equiv \alpha(w)\ ,\ \alpha(w) \in (\mathcal{B}^o)$$

Then

$$P\left[x(t,\alpha(w),w) = x(t,w)\right] = 1$$

Proof. We shall prove that

$$x(t,\alpha(w),w) = \alpha(w) + \int_0^t p(x(s,\alpha(w),w))d\beta(s,w) + \int_0^t r(x(s,\alpha(w),w))ds$$

with probability 1; then by uniqueness part of Theorem 1 the result will follow.

1. Since $x(t,a,w)$ is measurable in (a,w), $x(t,\alpha(w),w)$ is measurable in w. In fact, $x(t,\alpha(w),w)$ is the composite of

$$w \longrightarrow (\alpha(w),w) \quad \text{and} \quad (a,w) \longrightarrow x(t,a,w).$$

2. Consider the function-space valued random variable

$$\beta(w) = \beta(.,w) - \beta(o,w).$$

This induces a measure on $\mathbb{C}$, the space of all continuous functions on $[o,\infty)$ and with respect to this measure the set of coordinate functions is a Wiener process i.e. if for $\tilde{w} \in \mathbb{C}$

$$\tilde{\beta}(t,\tilde{w}) = \tilde{w}(t)$$

then $\tilde{\beta}(t,w)$ is a Wiener process on $\mathbb{C}$. Let $\tilde{\mathcal{B}}^t$ correspond to $\mathcal{B}^t$. There exists a unique solution of the equation

$$d\tilde{x}_t = p(\tilde{x}_t)\, d\tilde{\beta}(t) + r(\tilde{x}_t)dt \;, \quad \tilde{x}_o = a.$$

i.e. $$\tilde{x}(t,a,w) = a + \int_0^t p(\tilde{x}(s,a,\tilde{w}))d\tilde{\beta}(s,\tilde{w}) + \int_0^t r(\tilde{x}(s,a,\tilde{w}))ds$$

for almost all w. Hence we have by uniqueness

$$x(t,a,w) = \tilde{x}(t,a,\beta(w)) \quad \text{a.e.}$$

Let
$$L(a,w) = \tilde{x}(t,a,\tilde{w}) \quad \text{and}$$

$$R(a,\tilde{w}) = a + \int_0^t p(\tilde{x}(s,a,\tilde{w}))d\tilde{\beta}(s,\tilde{w}) + \int_0^t r(\tilde{x}(s,a,\tilde{w}))ds$$

If $\alpha(w) \in (\mathcal{B}^0)$ then $\alpha(w)$ and $\beta(w)$ are independent. Hence the measure induced by (α, β) on $R^1 \times \mathcal{C}$ is the product $P_\alpha \times P_\beta$ where P_α is the distribution of α and P_β is the probability induced on $\mathcal{C}$ by β. Hence we have

$$P\left[w: x(t,\alpha(w),w) = \alpha(w) + \int_0^t p(x(s,\alpha(w),w)d\beta(s,w) + \int_0^t r(x(s,\alpha(w),w)ds\right]$$

$$= (P_\alpha \times P_\beta)\left[(a,\tilde{w}) : L(a,\tilde{w}) = R(a,\tilde{w})\right]$$

$$= \int_R P_\beta\left[(a,\tilde{w}) : L(a,\tilde{w}) = R(a,\tilde{w})\right] P_\alpha(da) = \int_R 1 \cdot P_\alpha(da) = 1.$$

This proves the theorem.

§7. Construction of diffusion.

In this article, we shall answer the question of §1 i.e. we shall prove that if p and r satisfy Lipschitz condition then there exists a diffusion with state space R^1 such that if u, u', u'' are continuous, u and $\frac{1}{2}p^2u'' + ru'$ are bounded and $\mathcal{G}$ is the generator in the restricted sense, then

$$\mathcal{G}\, u = \tfrac{1}{2}p^2u'' + ru'.$$

We have proved in § 6 that

$$x(t) = a + \int_0^t p(x(s))d\beta(s) + \int_0^t r(x(s))ds$$

has a unique solution $x(t,a,w)$. Let $S = R^1$, $W = W_c(R^1)$ and

$$P_a(B) = P(w:x(.,a,w) \in B), \; B \in \mathcal{B}(W).$$

Then $\mathcal{M} = (S,W,P_a)$ is a diffusion.

We shall first prove that $\mathcal{M}$ is a Markov process. We verify the Markov property of P_a.

Let $\beta_t^-(w)$ denote the stopped path at t of $\beta(.,w)$ i.e. $\beta(t \wedge .,w)$ and let $\beta'(w) = \beta(t+.,w) - \beta(t,w)$, $\beta''(s,w) = \beta(t+s,w)$, $\mathcal{B}''^{\theta} = \mathcal{B}^{t+\theta}$ Then $\beta''(s,w)$ is also a Wiener process on Ω. Let $x(t,a,w)$, $y(t,a,w)$ denote solutions with respect to these processes of

$$dz_t = p(z_t)d\beta_t + r(z_t)dt$$

i.e.

$$x(t,a,w) = a + \int_0^t p(x(s,a,w))d\beta(s,w) + \int_0^t r(x(s,a,w))ds$$

$$y(t,b,w) = b + \int_0^t p(y(s.b,w))d\beta''(s,w) + \int_0^t r(y(s,b,w))ds$$

If $\beta(w) = \beta(.,w) - \beta(o,w)$ then $\beta(w)$ and $\beta'(w)$ induce the same probability on $\mathcal{C}$. Hence (see the proof of Theorem 3 of § 6)

$$x(t,a,w) = \tilde{x}(t,a,\beta(w)), \; y(t,a,w) = \tilde{x}(t,a,\beta'(w)).$$

Consider the $\mathcal{C}$-valued random variable $\beta_t^-(w)$. This induces a probability on $\mathcal{C}$ and with respect to this the process

$$\tilde{\beta}(s,\tilde{w}) = \tilde{w}(s) \quad , \quad 0 \le s \le t$$

is a Wiener process on $\mathcal{C}$ and there exists a unique solution for

$$d\tilde{x}_s = p(\tilde{x}_s)d\tilde{\beta}_s + r(\tilde{x}_s)ds \quad , \tilde{x}_0 = a \quad , \quad 0 \le s \le t$$

i.e. there exists $f(s,a,\tilde{w})$ such that

$$f(s,a,\tilde{w}) = a+\int_0^s p(f(\theta,a,\tilde{w}))d\tilde{\beta}(\theta,\tilde{w})+\int_0^s r(f(\theta,a,\tilde{w}))d\theta, \; 0 \le s \le t.$$

Then we have

$$f(s,a,\beta_t^-(w)) = a+\int_0^s p(f(\theta,a,\beta_t^-(w)))d\beta(\theta,w)+\int_0^s r(f(\theta,a,\beta_t^-(w)))d\theta,$$

$$0 \le s \le t.$$

Therefore the stopped path at t of $x(.,a,w)$ is

$$F(s,a,\beta_t^-(w)) = \begin{cases} f(s,a,\beta_t^-(w)) & , \quad 0 \le s \le t \\ f(t,a,\beta_t^-(w)) & , \quad s > t. \end{cases}$$

Now

$$x(t+s,a,w) = x(t,a,w)+\int_0^s p(x(\theta+t,a,w))d\beta(\theta+t,w)+ \int_0^s r(x(\theta+t,a,w))d\theta$$

$$= x(t,a,w)+\int_0^s p(x(\theta+t,a,w))d\beta''(\theta,w)+\int_0^s r(x(\theta+t,a,w))d\theta.$$

From Theorem 3 and uniqueness part of Theorem 1 of §6, we have therefore

$$x(t+s,a,w) = y(s,x(t,a,w),w) = \tilde{x}(s,x(t,a,w), \beta'(w))$$

$$= \tilde{x}(s,F(t,a,\beta_t^-(w)), \beta'(w))$$

Let $B_1 \in \mathcal{B}_t(W)$ and $B_2 \in \mathcal{B}(W)$. Then by definition of P_a

$$P_a\left[W \in B_1, W_t^+ \in B_2\right] = P\left[x(.,a,w) \in B_1 \; , \; x(t+.,a,w) \in B_2\right]$$

$$= P\left[F(.,a,\beta_t^-(w)) \in B_1' \; , \; \tilde{x}(.,F(t,a,\beta_t^-(w)), \; \beta'(w)) \in B_2\right]$$

where $B_1 = (w : w_t^- \in B_1')$. Let $P_{\beta_t^-}$ and $P_{\beta'}$ be the probabilities induced on $\mathcal{C}$ by β_t^- and β'; since they are independent they induce the product probability $P_{\beta_t^-} \times P_{\beta'}$ on $\mathcal{C} \times \mathcal{C}$. We have therefore,

$$P_a(w \in B_1, w_t^+ \in B_2) = (P_{\beta_t^-} \times P_{\beta'})\left[(\tilde{w},\tilde{w}'):F(.,a,\tilde{w}) \in B_1', \tilde{x}(.,F(t,a,\tilde{w}),\tilde{w}') \in B_2\right]$$

$$= \int P_{\beta_t^-}(d\tilde{w}) P\left[w':F(.,a.\tilde{w}) \in B_1', \tilde{x}(.,F(t,a,\tilde{w}),\tilde{w}') \in B_2\right]$$

$$= \int P_{\beta_t^-}(dw) P\left[w':F(.,a,\tilde{w}) \in B_1', \tilde{x}(.,F(t,a,\tilde{w}), \beta'(w')) \in B_2\right]$$

$$= \int P(dw) P\left[w':F(.,a,\beta_t^-(w)) \in B_1', \tilde{x}(.,F(t,a,\beta_t^-(w)), \beta'(w')) \in B_2\right]$$

$$= \int_{(w:F(.,a,\beta_t^-(w)) \in B_1')} P\left[w':\tilde{x}(.,F(t,a,\beta_t^-(w)), \beta'(w')) \in B_2\right] P(dw)$$

$$= \int_{(w:F(.,a,\beta_t^-(w)) \in B_1')} P\left[w':\tilde{x}(.,x(t,a,w), \beta(w')) \in B_2\right] P(dw)$$

since β and β' induce the same probability on $\mathcal{C}$. Thus by definition of P_b we have

$$P_a\left[w:w \in B_1, w_t^+ \in B_2\right] = \int_{(w:F(.,a,\beta_t^-(w)) \in B_1')} P_{x(t,a,w)}\left[B_2\right] P(dw)$$

$$= P_a\left[B_1 : P_{x_t(w)}(B_2)\right]$$

We have derived the Markov property.

From the remark at the end of Theorem 2 of §6, we see that if $a_n \longrightarrow a$, there exists a subsequence a_{nk} such that

$$x(t,a_{nk},w) \longrightarrow x(t,a,w) \quad \text{a.e.}$$

Since $H_t\, f(a) = E_a\,[f(x_t(W))] = \int_\Omega f(x(t,a,w))\; P(dw)$

if f is continuous and $a_n \longrightarrow a$ then there exists a subsequence a_{nk} such that $H_t\, f(a_{nk}) \longrightarrow H_t\, f(a)$. Since this is true of every sequence $a_n \longrightarrow a$, we should have

$$\lim_{b \to a} H_t\, f(b) = H_t\, f(a).$$

$\mathcal{M}$ is therefore a strong Markov process. The definition of P_a shows that $\mathcal{M}$ is conservative.

<u>Theorem 1</u>. If u, u', u'' are all continuous and if u and $\frac{1}{2}p^2u''+ru'$ are bounded, then $u \in \mathcal{D}(\mathcal{G})$ ($\mathcal{G}$ in the restricted sense) and

$$\mathcal{G} u = \tfrac{1}{2}\, p^2\, u'' + r\, u'.$$

<u>Proof</u>. It is enough to prove that $\alpha\, G_\alpha\, u - u = G_\alpha\, [\tfrac{1}{2}\, p^2\, u'' + r\, u']$. From the theorem of § 5, we have

$$u(x(t,a,w)) = u(x(o,a,w)) + \int_0^t u'(x(s,a,w))p(x(s,a,w))d\beta(s)$$
$$+ \int_0^t [\tfrac{1}{2}\, p^2(x(s,a,w))u''(x(s,a,w)) + u'(x(s,a,w))r(x(s,a,w))]\,ds.$$

Write $F(s,a,w) = \frac{1}{2}\, p^2(x(s,a,w))u''(x(s,a,w)) + u'(x(s,a,w))r(x(s,a,w))$.

Then since $x(o,a,w) = a$,

$$\int_{\Omega} u(x(t,a,w))P(dw) = u(a) + \int_0^t ds \int_{\Omega} F(s,a,w)P(dw) \qquad (1)$$

since the expectation of a stochastic integral is zero.

Thus

$$\alpha \int_0^{\infty} e^{-\alpha t} dt \int_{\Omega} u(x(t,a,w))dP(w) = u(a) + \int_0^{\infty} \alpha e^{-\alpha t} dt \int_0^t ds \int_{\Omega} F(s,a,w)dP(dw)$$

$$= u(a) + \int_0^{\infty} ds \int_{\Omega} F(s,a,w)P(dw) \int_s^{\infty} \alpha e^{-\alpha t} dt$$

$$= u(a) + \int_0^{\infty} ds \int_{\Omega} F(s,a,w)P(dw)e^{-\alpha s} \qquad \text{Q.E.D.}$$

<u>Theorem 2</u>. If u satisfies the conditions of Theorem 1, then

$$\lim_{t \to 0} \frac{H_t u(a) - u(a)}{t} = \tfrac{1}{2} p^2(a)\, u''(a) + r(a)\, u'(a).$$

This is immediate from equation (1) above, since all the functions involved are continuous.

<u>Theorem 3</u>. Let $P(t,a,E)$ be the transition probability of the above diffusion. Then the following Kolmogoroff conditions are true:

(A) $\displaystyle \lim_{t \to 0} \frac{1}{t} P(t,a,U_a^c) = 0$

(B) $\displaystyle \lim_{t \to 0} \frac{1}{t} \int_{U_a} (b-a)P(t,a,db) = r(a)$

(C) $\displaystyle \lim_{t \to 0} \frac{1}{t} \int_{U_a} (b-a)^2 P(t,a,db) = p^2(a)$

where U_a is any bounded open set containing a.

<u>Proof</u>. We can prove these facts using stochastic differential equations; but we shall deduce them from Theorem 2 above.

(A) Let V_a be any open set containing a with $\bar{V}_a \subset U_a$ and let u be a C_2 function such that $u = 0$ on V_a, $u = 1$ on U_a^c and $0 \leq u \leq 1$ on $U_a - V_a$. Then u satisfies the conditions of Theorem 1. We have

$$0 \leq \frac{1}{t} P(t,a,U_a^c) \leq \frac{1}{t}\left[H_t\, u(a)-u(a)\right] \longrightarrow \tfrac{1}{2} p^2 u''(a)+ru'(a)=0.$$

(B) Let $V_a \supset \bar{U}_a$ and let ϵ be a C_2 function vanishing outside V_a, 1 on $\bar{U}_a$ and $0 \leq \epsilon \leq 1$. Put $u(b) = (b-a)\,\epsilon(b)$. Then

$$\lim_{t\to 0} \frac{1}{t}\left[H_t\, u(a)-u(a)\right] = \tfrac{1}{2} p^2\, u''(a)+ru'(a) = r(a)$$

i.e.

$$\lim_{t\to 0} \frac{1}{t}\int_{R^1} u(b)\, P(t,a,db) = r(a).$$

Also

$$\overline{\lim_{t\to 0}}\, \left|\frac{1}{t}\int_{R^1} u(b)P(t,a,db)- \frac{1}{t}\int_{U_a} (b-a)P(t,a,db)\right|$$

$$\leq \overline{\lim_{t\to 0}}\, \frac{1}{t}\int_{V_a-U_a} |b-a|\,|\epsilon(b)-1|\,P(t,a,db) \leq \overline{\lim_{t\to 0}}\, C\, \frac{1}{t}\int_{U_a^c} P(t,a,db) = 0$$

from (A), where C is a bound for $|b-a|\,|\epsilon(b)-1|$ on V_a-U_a.

(C) Take $u(b) = (b-a)^2\epsilon(b)$ in (B).

<u>Remark</u>. Theorem 3 means (in an intuitive sense)

$$P_a(|dx_t| > \epsilon) = o(dt)\ ,\ E_a(dx_t) \sim r(a)dt\ ,$$

$$V_a(dx_t) = E_a((dx_t)^2) \sim p^2(a)dt.$$

Section 6

Linear diffusion

We recall the definition of a diffusion. A strong Markov process whose path functions are continuous before the killing time is called a diffusion. In this section we develop the theory (due to Feller) of linear diffusion.

§1. Generalities.

Definition. A diffusion whose state space S is a linear connected set is called a linear diffusion.

S is therefore one of the following sets, upto isomorphism i.e. order preserving homeomorphism of linear connected sets:

1) $[0,1]$, 2) $[0,1)$, 3) $(0,1]$, 4) $(0,1)$, 5) $\{0\}$.

Let σ_b denote the first passage time for b , i.e. $\sigma_b = \inf \{t : x_t = b\}$. If $P_a(\sigma_b < \infty) > 0$, we write $a \rightarrow b$; if $a \rightarrow b$ for some $b > a$, we write $a \in C_+$; if $a \rightarrow b$ for some $b < a$, we write $a \in C_-$. If $a \not\rightarrow b$ for any $b > a$, i.e. if $a \notin C_+$ we write $a \in K_-$; similarly if $a \not\rightarrow b$ for any $b \leq a$, i.e. if $a \notin C_-$, we write $a \in K_+$. Thus if $a \in K_+$ and $b < a$ then $P_a(\sigma_b = \infty) = 1$, i.e. $P_a(x_t \geqslant a$ for all $t < \sigma_\infty) = 1$.

Every point of the state space S belongs to one of the following sets:

1. $C_+ \cap C_- = K_+^c \cap K_-^c$. These points are called <u>regular points</u> or <u>second order points</u>.

2. $C_+ - C_- = K_+ - K_-$. A point of this set is called a <u>pure right shunt</u>.

3. $C_- - C_+ = K_- - K_+$. A point of this set is called a <u>pure left shunt</u>. Both left and right shunts are sometimes called <u>points of first order</u>, i.e. a point of first order is an element of $(C_+ - C_-) \cup (C_- - C_+)$.

4. $C_-^c \cap C_+^c = K_- \cap K_+$. These points are called <u>trap points</u> or <u>points of order zero</u>.

The intuitive meanings of the above should be clear; for instance a particle starting at a pure right shunt travels to the right with probability 1.

<u>Theorem 1</u>. If $a \in C_+ (\in C_-)$, then $P_a(\sigma_{a+} = 0) = 1$ $(P_a(\sigma_{a_-} = 0) = 1)$, where $\sigma_{a+} = \inf\{t : x_t > a\}$ $(\sigma_{a-} = \inf\{t : x_t < a\})$.

<u>Proof</u>. Let $\sigma = \sigma_{a_+}$ and $a \in C_+$. There exists $b > a$, and t such that $P_a(\sigma_b < t) > 0$. Now $E_a(e^{-\sigma_b}) \geqslant e^{-t} P_a(\sigma_b < t) > 0$. Since $\sigma_b(w) = \sigma(w) + \sigma_b(w_\sigma^+)$ we have, by the strong Markov property,

$$0 < E_a(e^{-\sigma_b}) = E_a(e^{-\sigma_b} : \sigma_b < \infty) = E_a(e^{-\sigma_b} : \sigma < \infty, \sigma_b < \infty)$$

$$= E_a(e^{-\sigma - \sigma_b(w_\sigma^+)} : \sigma < \infty, \sigma_b(w_\sigma^+) < \infty) = E_a\left[e^{-\sigma} E_{x_\sigma}(e^{-\sigma_b} : \sigma_b < \infty) : \sigma < \infty\right]$$

$$= E_a(e^{-\sigma} E_a(e^{-\sigma_b}) : \sigma < \infty) = E_a(e^{-\sigma_b}) E_a(e^{-\sigma}). \qquad \text{Q.E.D.}$$

<u>Remark</u>. If $\mathcal{M}$ is not strong Markov, the theorem is not true.

(e.g. exponential holding time process).

Theorem 2. If $a \longrightarrow b > a$ $(< a)$ then $[a,b) \subset C_+ ((b,a] \subset C_-)$.

Proof. Let $a \leq \xi < b$. Then $P_a(\sigma_\xi < \sigma_b) = 1$, because the paths are continuous. We have

$$0 < P_a(\sigma_b < \infty) = P_a(\sigma_\xi < \infty, \sigma_b < \infty) = P_a(\sigma_\xi < \infty, \sigma_b(w^+_{\sigma_\xi}) < \infty)$$

$$= E_a\left[P_{x_{\sigma_\xi}}(\sigma_b < \infty) : \sigma_\xi < \infty\right] = P_\xi\left[\sigma_b < \infty\right] P_a(\sigma_\xi < \infty),$$

since by continuity $x(\sigma_\xi) = \xi$. Therefore $P_\xi(\sigma_b < \infty) > 0$.
Q.E.D.

Corollary. If $a \in C_+$ then some right neighbourhood (i.e. a set which contains an interval $[a,b)$) of a, $U_+(a) \subset C_+$.

Theorem 3. The set of regular points is open.

Proof. Since $a \in C_+$, there exists $b > a$ with $P_a(\sigma_b < \infty) > 0$ and since $a \in C_-$, by Theorem 1, $P_a(\sigma_{a_-} = 0) = 1$. Hence $P_a(\sigma_{a_-} < \sigma_b < \infty) > 0$; this implies that there exists $c < a$ such that $P_a(\sigma_c < \sigma_b < \infty) > 0$. Noting that $a \in C_+$ and using Theorem 1, there exists d, $a < d < b$, with $P_a(\sigma_d < \sigma_c < \sigma_b < \infty) > 0$. Using the strong Markov property

$$P_a(\sigma_d < \infty) P_d(\sigma_c < \infty) P_c(\sigma_b < \infty) > 0,$$

so that $P_d(\sigma_c < \infty) > 0$, $P_c(\sigma_b < \infty) > 0$. Hence $(c,d] \subset C_-$ and $[c,b) \in C_+$. Q.E.D.

Theorem 4. K_+ is right closed, i.e. $a_n \in K_+$, $a_n \uparrow a$ imply $a \in K_+$ (K_- is left closed).

Proof. If $a \notin K_+$ then $a \in C_-$. There exists $b < a$ and $a \longrightarrow b$.

Then $(b,a] \subset C_-$ so that $(b,a] \cap K_+ = \phi$. Q.E.D.

§2. Generator in the restricted sense.

In the section of strong Markov processes we introduced a generator in the restricted sense; we modify this to suit our special requirements. Let $\mathcal{D}(S) = \{ f : f \in \mathcal{B}(S)$ and f is right continuous at every point of C_+ and left continuous at every point of $C_- \}$. $\mathcal{D}(S)$ is smaller than the classes $\mathcal{D}(S)$ introduced before in the section of strong Markov processes. Clearly $f \in \mathcal{D}(S)$ is continuous at every regular point and $\mathcal{D}(S) \supset C(S)$.

Theorem 1. $\mathcal{D}(S) \supset G_\alpha \mathcal{B}(S)$; a fortiori, $G_\alpha \mathcal{D}(S) \subset \mathcal{D}(S)$.

Proof. Let $a \in C_+$. Then , for $b > a$,

$$G_\alpha f(a) = E_a\left(\int_0^\infty e^{-\alpha t} f(x_t)dt\right)$$

$$= E_a\left(\int_0^{\sigma_b} e^{-\alpha t} f(x_t)dt\right) + E_a\left(\int_{\sigma_b}^\infty e^{-\alpha t} f(x_t)dt\right) = E_a\left(\int_0^{\sigma_b} e^{-\alpha t} f(x_t)dt\right) +$$

$$+ E_a\left(e^{-\alpha\sigma_b} G_\alpha f(x_{\sigma_b})\right)$$

$$= E_a\left(\int_0^{\sigma_b} e^{-\alpha t} f(x_t)dt\right) + E_a(e^{-\alpha\sigma_b}) G_\alpha f(b).$$

$$\text{Now } \left| E_a\left(\int_0^{\sigma_b} e^{-\alpha t} f(x_t)dt\right)\right| \leqslant \|f\| \frac{1-E(e^{-\alpha\sigma_b})}{\alpha} \xrightarrow[b \to a]{} \|f\| \frac{1-E(e^{-\alpha\sigma_{a+}})}{\alpha} = 0 ,$$

since $\qquad P_a(\sigma_{a_+} = 0) = 1.$ Q.E.D.

We can prove that $G_\alpha \mathcal{D}(S)$ is independent of α and the other results easily.

<u>Theorem 2</u>. $G_\alpha f = 0$, $f \in \mathcal{B}(S)$ imply $f \equiv 0$.

<u>Proof</u>. It is enough to show that $P_a(f(x_t) \longrightarrow f(a)$ as $t \longrightarrow 0)=1$.

If a is regular, f is continuous at a and there is nothing to prove. If $a \in C_+ - C_-$, f is right continuous at a and $P_a(x_t \geqslant a$ for $0 \leqslant t < \sigma_\infty) = 1$ and again the result is immediate. If $a \in C_- - \dot{C}_+$ the same is true. If a is a trap $P_a(x_t = a$ for $0 \leq t < \sigma_\infty) = 1$ and since $P_a(\sigma_\infty > 0) = 1$ (because $P_a(x_o = a) = 1$) the result follows again.

<u>Definition</u>. We define the generator in the restricted sense as $\mathcal{G} u = \alpha u - f$ where $u = G_\alpha f$ with $f \in \mathcal{B}(S)$. One easily verifies that $\mathcal{G} u$ is indep endent of α .

<u>Theorem 3</u>. If a is a trap, then $P_a(\sigma_\infty > t) \equiv P_a(\tau_a > t) = e^{-kt}$ and $\mathcal{G} u(a) = -k\, u(a)$ where $k \geqslant 0$ and τ_a = first leaving time from $a = \inf\{ t : x_t \neq a \}$.

<u>Proof</u>. Proceeding as in the case of a Markov process with discrete space (Section 2, §8) we show that $P_a(\tau_a > t) = e^{-kt}$ and $\frac{1}{k} = E_a(\tau_a)$ if $\infty > k > 0$. If $k = 0$, $P_a(\tau_a > t) = 1$ for all t , giving $P_a(\tau_a = \infty) = 1$ i.e. $P_a(x_t = a$ for all $t) = 1$ (such a point is called a <u>conservative trap</u>). We have, if $u = G_\alpha f$, $\mathcal{G} u(a) = \alpha u(a) - f(a) = \alpha \int_0^\infty e^{-\alpha t} E_a(f(x_t))dt - f(a) = f(a) - f(a) = 0$. Let now $\infty > k > 0$. Since $\frac{1}{k} = E_a(\tau_a)$, by Dynkin's formula,

$$E_a\left(\int_0^{\tau_a} \mathcal{G} u(x_t)dt\right) = E_a(u(x_{\tau_a})) - u(a) ,$$

i.e. $E_a(\tau_a \mathcal{G} u(a)) = -u(a)$, since $u(x_{\tau_a}) = u(\infty) = 0$. Q.E.D.

Theorem 4 (Dynkin). If a is not a trap then $E_a(\tau_U) < \infty$ for a sufficiently small open neighbourhood U of a and

$$\mathcal{G}\, u(a) = \lim_{U \to a} \frac{E_a(u(x_{\tau_U})) - u(a)}{E_a(\tau_U)},$$

where τ_U = first leaving time from U.

Proof. We prove that if a is not a trap, there exists $u_o \in \mathcal{D}(\mathcal{G})$ such that $\mathcal{G}u_o(a) > 0$. Let $\mathcal{G}\, u(a) = 0$ for every $u \in \mathcal{D}(\mathcal{G})$. Then for all $f \in C(S)$, $\alpha \cdot G_\alpha f(a) - f(a) = 0$ i.e. $\int_0^\infty H_t f(a)\, e^{-\alpha t} dt = \frac{1}{\alpha} f(a) = \int_0^\infty e^{-\alpha t} f(a) dt$. Since for $f \in C(S)$, $H_t f$ is right continuous in t, $H_t f(a) \equiv f(a)$ i.e. $\int f(b)\, P(t,a,db) = f(a)$ for all $f \in C(S)$. It follows that $P(t,a,db) = \delta_a(db)$ i.e. $P_a(x_t = a) = 1$ for all t. By right continuity, $P_a(x_t = a$ for all $t) = 1$, i.e. a is a trap. Thus there exists u_o such that $\mathcal{G}u_o(a) \neq 0$.

From the definition of $\mathcal{D}(S)$, we see that there exists $\epsilon_o > 0$ and a neighbourhood $U_o(a)$ such that

$$\mathcal{G}u_o(b) > \epsilon_o \begin{cases} \text{for } b \in U_o(a) \text{ if } a \text{ is regular,} \\ \text{for } b \in U_o(a) \text{ and } b \geq a \text{ if } a \text{ is a pure right shunt,} \\ \text{for } b \in U_o(a) \text{ and } b \leq a \text{ if } a \text{ is a pure left shunt.} \end{cases}$$

Therefore $P_a(\mathcal{G}u_o(x_t) > \epsilon_o$ for $0 \leq t < \tau_{U_o}) = 1$. Now put $\tau_n = n \wedge \tau_{U_o}$. Then

$$E_a\left(\int_0^{\tau_n} \mathcal{G}u_o(x_t) dt\right) = E_a(u_o(x_{\tau_n})) - u_o(a)$$

so that

$$\epsilon_o\, E_a(\tau_n) \leq 2\, \| u_o \| .$$

Letting $n \to \infty$, $E_a(\tau_{U_0}) \leq 2\frac{\|u_0\|}{\epsilon_0} < \infty$. Therefore for $U \subset U_0(a)$, $E_a(\tau_U) < \infty$.

Now let $u \in \mathcal{D}(\mathcal{G})$. For every $\epsilon > 0$, there exists an open neighbourhood $U(a) \subset U_0(a)$ such that

$$|\mathcal{G}u(b) - \mathcal{G}u(a)| < \epsilon \begin{cases} \text{for } b \in U(a) \text{ if } a \text{ is regular,} \\ \text{for } b \in U(a) \text{ and } b \geq a \text{ if } a \text{ is a pure right shunt,} \\ \text{for } b \in U(a) \text{ and } b \leq a \text{ if } a \text{ is a pure left shunt.} \end{cases}$$

Therefore $P_a(|\mathcal{G}u(x_t) - \mathcal{G}u(a)| < \epsilon \text{ for } 0 \leq t < \tau_U) = 1$. Using Dynkin's formula the proof can be easily completed.

§3. Local generator.

Let $\mathcal{M} = (S, W, P_a)$ denote a linear diffusion, and S' a closed interval in S Put $W' = W_c(S')$, $P'_a(B') = P_a(w^-_\tau \in B')$, where $\tau = \tau_{(S')^\circ}(w)$ is the first leaving time from the interior $(S')^\circ$ of S'. We prove that $\mathcal{M}' = (S', W', P'_a)$ is also a linear diffusion. We shall verify the strong Markov property for $\mathcal{M}'$. First we show that, if $\sigma'(w')$ is a Markov time in W', then $\sigma(w) = \sigma'(w^-_{\tau(w)})$ is a Markov time in W Now

$$(w: \sigma(w) \geq t) = [w: \sigma'(w^-_{\tau(w)}) \geq t)] = (w: w^-_{\tau(w)} \in B'_t), B'_t \in \mathcal{B}'_t$$

$$= (w: (w^-_{\tau(w)})^-_t \in B'), \; B' \in \mathcal{B}' \subset \mathcal{B}$$

$$= (w: t \leq \tau(w), w^-_t \in B') \cup (w: \tau(w) < t, (w^-_t)^-_{\tau(w)} \in B')$$

$$= (w: t \leq \tau(w), w^-_t \; B') \cup (w: \tau(w) < t, (w^-_t)^-_{\tau(w^-_t)} \in B') \in \mathcal{B}_t$$

since $w \longrightarrow w_t^-$ is $\mathcal{B}_t$-measurable and $w \longrightarrow w_{\sigma_1}^-$ is $\mathcal{B}$-measurable for any Markov time σ_1, we have $(w: (w_t^-)^-_{\sigma_1(w_t^-)} \in B) \in \mathcal{B}_t$ for any $B \in \mathcal{B}$.

Thus σ is a Markov time in W. Let $f_1' \in \mathcal{B}'_{\sigma'+}$ and $f_2' \in \mathcal{B}'$. Then by definition of P_a', we have

$$E_a'\left[f_1'(w')f_2'(w'^+_{\sigma'(w')})\right] = E_a\left[f_1'(w^-_{\tau(w)})f_2'((w^-_{\tau(w)})^+_{\sigma(w)})\right]$$

Put $f_1(w) = f_1'(w^-_{\tau(w)})$ and $f_2(w) = f_2^1(w^-_{\tau(w)})$. Let $\sigma_2 = \sigma \wedge \tau$. We show that $f_1 \in \mathcal{B}_{\sigma_2+}$. Now

$$f_1(w^-_{\sigma_2(w)+\delta}) = f_1'((w^-_{\sigma_2(w)+\delta})^-_{\tau(w^-_{\sigma_2(w)+\delta})}) = f_1'\left[(w^-_{\tau(w)})^-_{\sigma(w)+\delta}\right]$$

$$= f_1'\left[(w^-_{\tau(w)})^-_{\sigma'(w^-_{\tau(w)})+\delta}\right] = f_1'(w^-_{\tau(w)}), \text{ since } f_1' \in \mathcal{B}'_{\sigma'+}$$

This proves that $f_1 \in \mathcal{B}_{\sigma_2+}$. From the definition of τ and σ_2, we can see without difficulty that for any $t \geqslant 0$,

$$\sigma_2(w) + (t \wedge \tau(w^+_{\sigma_2(w)})) = \tau(w) \wedge (\sigma(w)+t)$$

Hence $(w^+_{\sigma_2(w)})^-_{\tau(w^+_{\sigma_2(w)})} = (w^-_{\tau(w)})^+_{\sigma(w)}$, so that

$$f_2(w^+_{\sigma_2(w)}) = f_2^1\left[(w^+_{\sigma_2(w)})^-_{\tau(w^+_{\sigma_2}(w))}\right] = f_2'\left[(w^-_{\tau(w)})^+_{\sigma(w)}\right].$$

Thus

$$E'_a\left[f'_1(w')f'_2(w'^+_{\sigma'(w')})\right] = E_a\left[f'_1(w^-_{\tau(w)})f'_2((w^-_{\tau(w)})^+_{\sigma(w)})\right]$$

$$= E_a\left[f_1(w)f_2(w^+_{\sigma_2})\right] = E_a\left[f_1(w)E_{x_{\sigma_2}}(f_2(w))\right]$$

$$= E_a\left[f'_1(w^-_{\tau(w)})E_{x_{\sigma(w)}(w^-_{\tau(w)})}(f'_2(w^-_{\tau(w)}))\right] = E'_a\left[f'_1(w')E_{x_{\sigma'}}(f'_2(w'))\right]$$

which proves that $\mathbb{M}'$ is a linear diffusion. $\mathbb{M}'$ is called the <u>stopped process</u> at the boundary $\partial S'$ of S'. We also denote $\mathbb{M}'$ by $\mathbb{M}_{S'}$, its generator by $\mathcal{G}$ or $\mathcal{G}_{S'}$ etc.

A point $a \in S$ is called a <u>conservative point</u> if there exists a neighbourhood U of a such that $\mathbb{M}_{\bar{U}}$ is conservative. The set of all conservative points is evidently open. Let a be a conservative regular point and S' a closed interval containing a such that $\mathbb{M}_{S'}$ is conservative. We shall prove that if $u \in \mathcal{D}(\mathcal{G})$, then $u' = u/S' \in \mathcal{D}(\mathcal{G}')$ and $\mathcal{G}'u' = \mathcal{G}u$ in $(S')^\circ$; more generally if $S' \supset S''$, if $u' \in \mathcal{D}(\mathcal{G}_{S'})$ then $u'' = u'/S''$ (restriction to S'') is in $\mathcal{D}(\mathcal{G}_{S''})$ and $\mathcal{G}'u' = \mathcal{G}''u''$ in $(S'')^\circ$. Then we can define $\mathcal{G}_a$ the local generator as the inductive limit of $\mathcal{G}_{S'}$ as $S' \downarrow a$ in the following way. Consider the set $\mathcal{D}_a$ of all functions defined in a neighbourhood of a (which may depend on the function) right (left) continuous at points of C_+ (C_-). Introduce an equivalence relation in $\mathcal{D}_a$ by putting $f \sim g$ if and only if there exists a neighbourhood U of a such that $f = g$ in U. Let $\overline{\mathcal{D}}_a(S) = \mathcal{D}_a(S)/\sim$ (the equivalence classes). Define $\mathcal{D}(\mathcal{G}_a) = \{\bar{u} : \bar{u} \in \overline{\mathcal{D}_a}(S)$ and there exists $U = U(a)$ with $u/U \in \mathcal{D}(\mathcal{G}_{\bar{U}})\}$.

Define $\mathcal{G}_a \bar{u} = (\mathcal{G}_{\bar{U}} u)/\sim$ where $\bar{u} = u/\sim$, $u|U \in \mathcal{D}(\mathcal{G}_{\bar{U}})$. From above, it follows that this is independent of the choice of U. We now prove that if $u \in \mathcal{D}(\mathcal{G})$ then $u' = u|S' \in \mathcal{D}(\mathcal{G}')$ and $\mathcal{G}u = \mathcal{G}'u'$ in $(S')^\circ$. Note that if $[b,c] = S'$, $\tau = \tau_U = \sigma_b \wedge \sigma_c$, $U = (S')^\circ$. We have

$$u(\xi) = G_\alpha f(\xi) = E_\xi\left(\int_0^\infty e^{-\alpha t} f(x_t)dt\right)$$

$$= E_\xi\left(\int_0^\tau e^{-\alpha t} f(x_t)dt\right) + E_\xi\left(\int_{\sigma_b}^\infty e^{-\alpha t} f(x_t)dt : \sigma_b < \sigma_c\right) + E_\xi\left(\int_{\sigma_c}^\infty e^{-\alpha t} f(x_t)dt : \sigma_c < \sigma_b\right)$$

$$= E_\xi\left(\int_0^\tau e^{-\alpha t} f(x_t(w_\tau^-))dt\right) + G_\alpha f(b) E_\xi\left(e^{-\alpha\sigma_b} : \sigma_b < \sigma_c\right) + G_\alpha f(c) E_\xi\left(e^{-\alpha\sigma_c} : \sigma_c < \sigma_b\right)$$

by strong Markov property. Put $f' = f$ in U, $f'(b) = \alpha G_\alpha f(b)$ and $f'(c) = \alpha G_\alpha f(c)$. Then it is easy to show that $u' = u|S' = G'_\alpha f'$ and $\mathcal{G}'u' = \mathcal{G}u$ in U.

<u>Definition</u>. $\mathcal{G}_a$ is called the <u>local generator</u> at a.

§4. <u>Feller's form of generators</u> (1) - <u>Scale</u>

We shall derive Feller's canonical form of generators by purely probabilistic methods following Dynkin in the following articles.

Let a be a conservative regular point. There exists $U = U(a) = (b,c)$ such that $\bar{U}$ has only conservative regular points and $E_\xi(\tau_U) = E_\xi(\sigma_{\partial U}) < \infty$ for $\xi \in U$. Put $s(\xi) = P_\xi(\sigma_c < \sigma_b)$.

(1°) $s \in \mathcal{D}(\mathcal{G}_{\bar{U}})$ and $\mathcal{G}_{\bar{U}}\, s = 0$ in $\bar{U}$.

Let $f(c) = 1$ and $f(\xi) = 0$, $\xi \in [b,c)$. Then $f \in \mathcal{D}(\mathcal{G}_{\bar{U}}) = \mathcal{D}'$ and

$$G'_\varepsilon f(\xi) = E'_\xi\left(\int_0^\infty e^{-\varepsilon t} f(x_t)dt\right) = E_\xi\left(\int_0^\infty e^{-\varepsilon t} f(x_t(w^-_{\tau_U}))\,dt\right) =$$

$$= E_\xi\left(\int_{\sigma_c}^\infty e^{-\varepsilon t}\,dt : \sigma_c < \sigma_b\right).$$

Hence $\lim_{\varepsilon \downarrow 0} \varepsilon G'_\varepsilon f(\xi) = P_\xi(\sigma_c < \sigma_b) = s(\xi)$. The resolvent equation gives

$$(G'_\alpha - G'_\varepsilon) f + (\alpha - \varepsilon) G'_\alpha G'_\varepsilon f = 0 \quad \text{or}$$

$$\varepsilon (G'_\alpha - G'_\varepsilon) f + (\alpha - \varepsilon) G'_\alpha G'_\varepsilon f = 0 .$$

Letting $\varepsilon \to 0$, we get $-s(\xi) + \alpha G'_\alpha s(\xi) = 0$. Therefore, firstly $s \in \mathcal{D}'$ and again, since $s = \alpha G'_\alpha s$, $s \in \mathcal{D}(\mathcal{G}')$ and

$$\mathcal{G}' s = \alpha s - (G'_\alpha)^{-1} s = \alpha s - \alpha s = 0 \quad \text{in } \overline{U} .$$

(2°) $s(\xi)$ is continuous in $\overline{U}$.

Since $s \in \mathcal{D}(\mathcal{G}')$ and all points of U are regular, s is continuous in U.

It remains to prove that s is continuous at b and c. We prove the continuity at c; continuity at b is proved in the same way. To prove this, we shall first prove that

$e = \lim_{\xi \uparrow c} E_\xi(e^{-\sigma_{c-}}) = 1$ or 0, $\sigma_{c-} = \lim_{\eta \uparrow c} \sigma_\eta$. Let $\xi < \eta < \zeta < c$.

Then $E_\xi(e^{-\sigma_\zeta}) = E_\xi(e^{-\sigma_\eta})E_\eta(e^{-\sigma_\zeta})$. Letting $\zeta \uparrow c$,

$E_\xi(e^{-\sigma_{c-}}) = E_\xi(e^{-\sigma_\eta})E_\eta(e^{-\sigma_{c-}})$. Letting $\eta \uparrow c$ now and $\xi \uparrow c$ finally, we get $e = e^2$ so that $e = 1$ or 0. Since c is regular, there exists $\xi < c$ such that $P_\xi(\sigma_c < \infty) > 0$ and then

$E_\xi(e^{-\sigma_c}) > 0$. Also $\sigma_c \geq \sigma_{c-}$. It follows that $E_\xi(e^{-\sigma_{c-}}) > 0$. Hence $e = 1$. Since ξ is conservative,

$$P_\xi(x_{\sigma_{c-}} = \infty,\ \sigma_{c-} < \infty) \leq P_\xi(\sigma_\infty < \infty) = 0.$$

Therefore, since $\sigma_c \geq \sigma_{c-}$ and the paths are continuous before the killing time, $P_\xi(\sigma_c = \sigma_{c-}) = 1$. We have proved that $\lim\limits_{\xi \uparrow c} E_\xi(e^{-\sigma_c}) = 1$. For every $\varepsilon > 0$, therefore $\lim\limits_{\xi \to c} P_\xi(\sigma_c < \varepsilon) = 1$. Also

$$s(\xi) = P_\xi(\sigma_c < \sigma_b) \geq P_\xi(\sigma_c < \varepsilon, \sigma_b \geq \varepsilon) \geq P_\xi(\sigma_c < \varepsilon) - P_\xi(\sigma_b < \varepsilon).$$

If $b < \xi_0 < \xi < c$, then

$$P_\xi(\sigma_b < \varepsilon) \leq P_\xi(\sigma_{\xi_0} < \infty, \sigma_b(w^+_{\sigma_{\xi_0}}) < \varepsilon) = P_\xi(\sigma_{\xi_0} < \infty)\, P_{\xi_0}(\sigma_b < \varepsilon) \leq P_{\xi_0}(\sigma_b < \varepsilon).$$

Therefore $s(\xi) \geq P_\xi(\sigma_c < \varepsilon) - P_{\xi_0}(\sigma_b < \varepsilon)$. Letting $\xi \uparrow c$ first and $\varepsilon \downarrow 0$ next, we get $\lim\limits_{\xi \to c} s(\xi) \geq 1$ i.e. $s(\xi)$ is continuous at $\xi = c$.

(3^0) $s(\xi)$ is strictly increasing.

The set of points ξ, $b < \xi \leq c$ such that $s(\xi) = 0$ is closed in $(b,c]$. If $P_{\xi_0}(\sigma_c < \sigma_b) = 0$, the same is evidently true for any $b < \xi < \xi_0$. Since ξ_0 is regular, $\lim\limits_{\eta \downarrow \xi_0} P_{\xi_0}(\sigma_\eta < \varepsilon) = 1$ for any $\varepsilon > 0$. Also $P_{\xi_0}(\sigma_b > 0) = 1$. It easily follows that $\lim\limits_{\eta \downarrow \xi_0} P_{\xi_0}(\sigma_\eta < \sigma_b) = 1$. Choose $\eta_0 > \xi_0$ with $P_{\xi_0}(\sigma_{\eta_0} < \sigma_b) > 0$. Then $P_{\xi_0}(\sigma_\eta < \sigma_b) > 0$ for any $\xi_0 < \eta < \eta_0$. Now note that if $a < \xi$, then $(\sigma_a < \sigma_\xi) = (w: \sigma_\xi(w^-_{\sigma_a}) = \infty)$, and hence is in $\mathcal{B}_{\sigma_{\bar a}}$

We have $0 = P_{\xi_0}(\sigma_c < \sigma_b) = P_{\xi_0}(\sigma_\eta < \sigma_b)P_\eta(\sigma_c < \sigma_b)$. Thus $P_\eta(\sigma_c < \sigma_b) = 0$. The connectedness of $(b,c]$ shows that $s(\xi) \neq 0$ in $(b,c]$. Exactly similar argument also shows that $s(\xi) < 1$ in $[b,c)$. Now if $\xi < \eta$, we replace c by η and repeat the argument to get $P_\xi(\sigma_\eta < \sigma_b) < 1$. Thus if $\xi < \eta$,

$$s(\xi) = P_\xi(\sigma_c < \sigma_b) = P_\xi(\sigma_\eta < \sigma_b)P_\eta(\sigma_c < \sigma_b) < P_\eta(\sigma_c < \sigma_b) = s(\eta).$$

(4°) $\alpha s + \beta$ is the general solution of $\mathcal{G}'u = 0$.

Let $f(\xi) = 1$ for $b \leq \xi \leq c$. Then $f(\xi) = 1 = \alpha E_\xi(\int_0^\infty e^{-\alpha t} f(x_t)dt) = \alpha G'_\alpha f(\xi)$. This firstly shows that $f \in \mathcal{D}(\mathcal{G}')$ and then the same equation shows that $f \in \mathcal{D}(\mathcal{G}')$. Thus $\mathcal{G}'1 = \alpha \cdot 1 - (G'_\alpha)^{-1} 1 = \alpha - \alpha = 0$. Hence since $\mathcal{G}'s = 0$ $\mathcal{G}'(\alpha s + \beta) = 0$. Now let $\mathcal{G}'u = 0$. Then $0 = E'_\xi(\int_0^{\tau_U} \mathcal{G}'u(x_t)dt)$ $= E'_\xi(u(x_{\tau_U})) - u(\xi) = E_\xi(u(x_{\tau_U})) - u(\xi)$

$$= u(b)\, P_\xi(\sigma_b < \sigma_c) + u(c) P_\xi(\sigma_c < \sigma_b) - u(\xi).$$

Therefore u is linear in s.

(5°) If $b < b' \leq \xi \leq c' < c$ then $P_\xi(\sigma_{c'} < \sigma_{b'}) = \dfrac{s(\xi) - s(b')}{s(c') - s(b')}$

Let $x = P_\xi(\sigma_{c'} < \sigma_{b'})$, $y = P_\xi(\sigma_{b'} < \sigma_{c'})$; then $x + y = 1$ and

$$P_\xi(\sigma_c < \sigma_b) = P_\xi(\sigma_{c'} < \sigma_b) P_{c'}(\sigma_c < \sigma_b)$$

Also $(\sigma_{c'} < \sigma_b) = (\sigma_{c'} < \sigma_{b'}) \cup (\sigma_{c'} > \sigma_{b'}, \sigma_b(w^+_{\sigma_{b'}}) > \sigma_{c'}(w^+_{\sigma_{b'}}))$

Therefore $P_\xi(\sigma_c<\sigma_b) = P_\xi(\sigma_{c'}<\sigma_{b'})P_{c'}(\sigma_c<\sigma_b)$ +

$$+ P_\xi(\sigma_{c'}>\sigma_{b'},\ \sigma_b(w^+_{\sigma_{b'}})>\sigma_{c'}(w^+_{\sigma_{b'}}))P_{c'}(\sigma_c<\sigma_b)$$

$$= x\ s(c')+P_\xi(\sigma_{c'}>\sigma_{b'})P_{b'}(\sigma_b>\sigma_{c'})P_{c'}(\sigma_c<\sigma_b)$$

$$= x\ s(c')+P_\xi(\sigma_{b'}<\sigma_{c'})P_{b'}(\sigma_c<\sigma_b)$$

i.e. $s(\xi) = x\ s(c')+y\ s(b')$. Solving for x we get the result.

Definition. s is called the canonical scale in $[b,c]$.

§5. Feller's form of generators (2) - Speed measure.

Let $p(\xi) = E'_\xi(\tau_U) = E_\xi(\tau_U)$, $U = (b,c)$. Put $f(x) = 1$ for $x \in U$, $f(b) = f(c) = 0$. Then $G'_\epsilon f(\xi) = E'_\xi(\int_0^\infty e^{-\epsilon t} f(x_t)dt) = E_\xi(\int_0^{\tau_U} e^{-\epsilon t}dt)$, so that

$$\lim_{\epsilon\downarrow 0} G'_\epsilon f(\xi) = p(\xi)$$

We have $G'_\alpha f - G'_\epsilon f + (\alpha-\epsilon)G'_\alpha G'_\epsilon f = 0$. Letting $\epsilon \longrightarrow 0$

$$G'_\alpha f - p + \alpha G'_\alpha p = 0.$$

This shows that $p \in \mathcal{D}(\mathcal{G}')$, because, f being identically 1 in U, is continuous at every regular point and b,c are traps for M'.

We have,

(1°) $\mathcal{G}'p = -f$ i.e. $\mathcal{G}'p = -1$ in U, $\mathcal{G}'p(b) = \mathcal{G}'p(c) = 0$

(2°) p is continuous in $\bar{U}$ and $p(b) = p(c) = 0$.

We prove that $p(c-) = p(c) = 0$. Let $b < \xi < c$ and

$\tau_\xi = \tau_{(b,\xi)}$. Then if $b < \xi_0 < \xi$,

$$E_{\xi_0}(\tau_U) = E_{\xi_0}(\tau_\xi) + E_{\xi_0}(\tau_U(w^+_{\tau_\xi})) = E_{\xi_0}(\tau_\xi) + E_{\xi_0}(E_{x_{\tau_\xi}}(\tau_U) : \sigma_\xi < \sigma_b)$$

$$= E_{\xi_0}(\tau_\xi) + E_\xi(\tau_U)\, P_{\xi_0}(\sigma_\xi < \sigma_b).$$

Now as $\xi \to c$, $E_{\xi_0}(\tau_\xi) \longrightarrow E_{\xi_0}(\tau_U)$ and $\sigma_\xi \to \sigma_c$. Therefore

$\lim_{\xi \to c} E_\xi(\tau_U)\, P_{\xi_0}(\sigma_c < \sigma_b) = 0$ i.e. $p(c-) = 0$.

(3^0) p is the only solution of $\mathcal{G}'u = -1$ in U, $u(b) = u(c) = 0$.

For, $-p(\xi) = -E_\xi(\tau_U) = E_\xi\left(\int_0^{\tau_U} \mathcal{G}'u(x_t)dt\right) = E_\xi(u(x_{\tau_U})) - u(\xi)$.

Since $x_{\tau_U} = b$ or c, $u(x_{\tau_U}) = 0$. Q.E.D.

(4^0) We have proved that $s: [b,c] \to [0,1]$ is 1-1 continuous.

We define a mapping p' on $[0,1]$ by $p'(s(\xi)) = p(\xi)$. To prove that p' is strictly concave in $[0,1]$, we express this by "p is concave in s". We have to prove that, if $b \le \eta < \xi < \zeta \le c$,

$$p(\xi) > \frac{s(\xi)-s(\eta)}{s(\zeta)-s(\eta)}\, p(\zeta) + \frac{s(\zeta)-s(\xi)}{s(\zeta)-s(\eta)}\, p(\eta).$$

Now $p(\xi) = E_\xi(\tau_U) = E_\xi(\sigma + \tau_U(w^+_\sigma))$, $\sigma = \tau_{(\eta,\zeta)}$,

$$> E_\xi(\tau_U(w^+_\sigma)) = \text{right side of the above inequality.}$$

(5^0) $m(\xi) = \dfrac{d^+p}{ds}$ is strictly increasing and bounded.

(The measure dm is called the <u>speed measure</u> for $\bar{U}$).

From (4^0), the right derivative $\dfrac{d^+p}{ds}$ exists and strictly increases. We prove that it is bounded. Let $V = (b_1, c_1) \supset [b,c]$. Put $p_1(\xi) = E_\xi(\tau_V)$, $s_1(\xi) = P_\xi(\sigma_{c_1} < \sigma_{b_1})$. We have

$$p_1(\xi) = E_\xi(\tau_U) + E_\xi(\tau_V(w^+_{\tau_U})) = p(\xi) + s(\xi)p_1(c) + (1-s(\xi))p_1(b).$$

From this one easily sees that

$$m_1(\xi) = -\frac{d^+p_1(\xi)}{ds_1} = [m(\xi) - (p_1(c) - p_1(b))]\frac{1}{s_1(c) - s_1(b)}$$

Q.E.D.

§6. <u>Feller's form of generators</u> (3).

<u>Theorem (Feller)</u>. $u \in \mathcal{D}(\mathcal{G}')$ if and only if

(1) u is of bounded variation in U

(2) $du < ds$ i.e. du is absolutely continuous with respect to ds.

(3) $\frac{du}{ds}$ (Radon Nikodym derivative) is of bounded variation in U.

(4) $d\frac{du}{ds} < dm$ in U.

(5) $(d\frac{du}{ds})/dm$ (which we shall write $\frac{d}{dm}\frac{du}{ds}$) has a continuous version in U.

(6) u is continuous at b and c i.e. u is continuous in $\bar{U}$

and $\mathcal{G}'u = \frac{d}{dm}\frac{du}{ds}$ in U, $\mathcal{G}'u = 0$ at b and c.

<u>Proof</u>. (Dynkin) Let $u \in \mathcal{D}(\mathcal{G}')$. Then for some $f \in \mathcal{D}'$

$$u(\xi) = G'_\alpha f(\xi) = E_\xi\left(\int_0^{\tau_U} e^{-\alpha t} f(x_t)dt\right) + E_\xi(e^{-\alpha\sigma_b} : \sigma_b < \sigma_c)\frac{f(b)}{\alpha}$$
$$+ \frac{f(c)}{\alpha} E_\xi(e^{-\alpha\sigma_c} : \sigma_c < \sigma_b).$$

Thus $\lim_{\xi \to b} u(\xi) = \frac{f(b)}{\alpha} = u(b)$ and $\lim_{\xi \to c} u(\xi) = \frac{f(c)}{\alpha} = u(c)$. u is therefore continuous in $\bar{U}$.

Let $[\alpha,\beta]\subset U$. If $\mathcal{G}'v \geq 0$ in (α,β) then Dynkin's formula shows, for $\alpha < \xi < \beta$,

$$0 \leq E_\xi\left(\int_0^{\sigma_\alpha\wedge\sigma_\beta} \mathcal{G}'v(x_t)dt\right) = v(\alpha)\frac{s(\beta)-s(\xi)}{s(\beta)-s(\alpha)} - v(\beta)\frac{s(\xi)-s(\alpha)}{s(\beta)-s(\alpha)} - v(\xi) ,$$

so that v is convex in s and hence is of bounded variation in $[\alpha,\beta]$. Also $\frac{d^+v}{ds}$ exists and increases in $[\alpha,\beta]$ and dv is absolutely continuous with respect to ds. If $\mathcal{G}'v \geq \lambda$ in (α,β), then $\mathcal{G}'(v+\lambda p) \geq 0$ in (α,β). Therefore $d\frac{d^+v}{ds} \geq \lambda\, dm$. Similarly if $\mathcal{G}'v \leq \mu$ in (α,β) then $d\frac{d^+v}{ds} \leq \mu\, dm$.

Consider a division $\Delta = (b=\alpha_0<\alpha_1<\ldots<\alpha_n = c)$ of $[b,c]$. Put

$$\lambda_i = \inf_{\xi\in(\alpha_i,\alpha_{i+1})} \mathcal{G}'u(\xi) \quad , \mu_i = \sup_{\xi\in(\alpha_i,\alpha_{i+1})} \mathcal{G}'u(\xi).$$

Then $\mu_i\, dm \geq d\frac{d^+u}{ds} \geq \lambda_i\, dm$ in (α_i,α_{i+1}) and $\mu_i\, dm \geq \mathcal{G}'u\, dm \geq \lambda_i\, dm$ in (α_i,α_{i+1}) Putting $\lambda(\xi) = \lambda_i$ and $\mu(\xi)=\mu_i$ for $\alpha_i \leq \xi < \alpha_{i+1}$, we have

$$\mu(\xi)dm \geq d\frac{d^+u}{ds} \geq \lambda(\xi)dm , \text{ and } \mu(\xi)dm \geq \mathcal{G}'u(\xi)dm \geq \lambda(\xi)dm.$$

Therefore $(\mu(\xi)-\lambda(\xi))dm \geq d\frac{d^+u}{ds} - \mathcal{G}'u(\xi)dm \geq -(\mu(\xi)-\lambda(\xi))dm$.

As $\delta(\Delta) = \max_i [\alpha_{i+1}-\alpha_i]$ tends to zero, $\mu(\xi)-\lambda(\xi) \longrightarrow 0$.

We have

$$d\frac{d^+u}{ds} = \mathcal{G}' u \, dm \quad \text{in } U.$$

Conversely suppose that u satisfies all the above six conditions. Define $f = \alpha u - \frac{d}{dm}\frac{d}{ds}u$ in U and $f(b) = \alpha u(b)$, $f(c) = \alpha u(c)$. Then since f is continuous in U, $f \in \mathcal{D}'$. Let $v = G'_\alpha f$. From what we have already proved, $\alpha v - \frac{d}{dm}\frac{d}{ds}v = f$ in U, $v(b) = \frac{1}{\alpha}f(b)$, $v(c) = \frac{1}{\alpha}f(c)$. If $\theta = u-v$, then θ is continuous in $\bar{U}$, $\theta(b) = \theta(c) = 0$, and $\alpha\theta - \frac{d}{dm}\frac{d}{ds}\theta \equiv 0$. There exists a point ξ_0 such that $\theta(\xi_0)$ is a maximum. Now $\theta(\xi_0) > 0 \Longrightarrow \theta(\xi) > 0$ near $\xi_0 \Longrightarrow \frac{d}{dm}\frac{d}{ds}\theta > 0$ near $\xi_0 \Longrightarrow \frac{d}{ds}\theta$ strictly increases near ξ_0. Then if $\xi > \xi_0 > \eta$ are near ξ_0 we have $\theta(\xi_0) - \theta(\eta) = \int_\eta^{\xi_0} \frac{d\theta}{ds}ds < \frac{d\theta}{ds}(\xi_0)[s(\xi_0) - s(\eta)]$ Hence $\frac{d\theta}{ds}(\xi_0) > 0$. On the other hand $\theta(\xi) - \theta(\xi_0) = \int_{\xi_0}^{\xi} \frac{d\theta}{ds}ds > \frac{d\theta}{ds}(\xi_0)[s(\xi) - s(\xi_0)]$, a contradiction. Therefore $\theta(\xi) \leq 0$. Similarly we prove $\theta(\xi) \geq 0$. Q.E.D.

§7. <u>Feller's form of generators</u> (4) - <u>Conservative compact interval</u>.

Let $I = [b,c]$ be a conservative compact regular interval i.e. a compact interval consisting only of conservative regular points. We shall prove the following

<u>Theorem</u> (Feller). All the results of the last three articles hold for M_I.

<u>Proof</u>. Since every $a \in I$ is conservative regular, we can associate with any $a \in I$ an open interval $U(a)$ such that $E_\xi(\tau_{U(a)}) < \infty$ for $\xi \in U(a)$ and then the results of the last three articles are

true for $\mathcal{M}_{\overline{U(a)}}$. Denote the quantities s,p,m etc. for $\mathcal{M}_{\bar{U}}$ by s_U, p_U, m_U , etc. Let $s = P_\xi(\sigma_c < \sigma_b)$. Then from ($5^0$) of § 4, we get

$$s(\xi) = s(b') + [s(c')-s(b')] P_\xi(\sigma_{c'} < \sigma_{b'}),$$

where $\xi \in (b',c')$ is an interval such that the results of the last three articles are true for $\mathcal{M}_{[b',c']}$. This equation shows that $s(\xi)$ is strictly increasing and continuous in some neighbourhood of the point ξ . Therefore $s(\xi)$ is strictly increasing and continuous in I , and s is linear in s_U in U , for every interval U such that the results of the last three articles are true for $\mathcal{M}_{\bar{U}}$. Let dm be a measure defined on $\mathcal{B}(I)$ as follows. $dm = \frac{1}{\alpha_U} dm_U$ if in U , $s = \alpha_U s_U + \beta_U$. Let $V \cap U = W \neq \emptyset$. Since $p_U = p_W + p_U(b') + s_W(\xi)[p_U(c')-p_U(b')]$ where $U = (b',c')$ we have $\frac{1}{\alpha_U} dm_U = dm_W = \frac{1}{\alpha_V} dm_V$. Therefore the measure dm is uniquely defined on $\mathcal{B}(I)$ and $\frac{d}{dm}\frac{d}{ds} = \frac{d}{dm_U}\frac{d}{ds_U}$ in U. dm is defined by a strictly increasing function m (say) in I. Consider now the following "differential equation"

$$\frac{d}{dm}\frac{d}{ds} u = -1 \text{ in } (b,c) \text{ and } u(b+) = u(c-) = 0.$$

Then $p(\xi) = -\int_{b+}^{\xi} m(\eta)ds(\eta) + \int_{b+}^{c-} m(\eta)d\,s(\eta)\,[s(\xi)-s(b)]\,\frac{1}{s(c)-s(b)}$

is a solution and $(\alpha - \frac{d}{dm}\frac{d}{ds})p = \alpha p+1$ in (b,c) and $p(b+) = p(c-) = 0$. Let $f = \alpha p+1$ in (b,c) and $f(b) = f(c) = 0$. Since f is continuous in (b,c), $f \in \mathcal{D}_I$. Let $v = G^I f$. Then $v \in \mathcal{D}(\mathcal{G}^I)$

and $(\alpha - \mathcal{G}^I)v = \alpha p + 1$. $v \in \mathcal{D}(\mathcal{G}^I) \Rightarrow v \in \mathcal{D}(\mathcal{G}^{\bar{U}})$ so that $(\alpha - \frac{d}{dm_U}\frac{d}{dS_U})v = \alpha p + 1$ in U. Therefore $(\alpha - \frac{d}{dm}\frac{d}{ds})v = \alpha p + 1$ in (b,c). Since $v \in \mathcal{D}(\mathcal{G}^I)$, it is continuous in I. Let $\theta = p - v$. θ is continuous in I and $(\alpha - \frac{d}{dm}\frac{d}{ds})\theta = 0$. We prove as in §6 that $\theta = 0$. Thus $p(\xi) = v(\xi) \in \mathcal{D}(\mathcal{G}^I)$. Using Dynkin's formula we have, if $\tau_n = \tau_{(b,c)} \wedge n = \tau \wedge n$ (say)

$$E_\xi\left(\int_0^{\tau_n} \mathcal{G}^I(p(x_t)dt\right) = E_\xi(p(x_{\tau_n})) - p(\xi)$$

i.e. $E_\xi(\tau_n) \leq 2\|p\| < \infty$. We get $E_\xi(\tau) < \infty$.

Again using Dynkin's formula

$$E_\xi\left(\int_0^{\tau} \mathcal{G}^I p(x_t)dt\right) = E_\xi(p(x_\tau)) - p(\xi) \text{ i.e. } p(\xi) = E_\xi(\tau_{(b,c)}).$$

The proof of the theorem can be completed as in §6.

References

Section 0

DOOB, J.L. - Stochastic Processes, New York (1952)

HALMOS, P.R. - Measure Theory, New York (1950)

KOLMOGOROV, A.N. - Grundbegriffe der Wahrscheinlichkeitsrechnung Ergeb. der Math. 2, 3 (1933) [English translation: Foundations of the theory of Probability (1956)]

Sections 1 and 2

BLUMENTHAL, R.M. - An extended Markov property, Trans. Amer. Math. Soc., 85(1957), pp.52-72

CHUNG, K.L. - On a basic property of Markov chains, Ann. Math. 68(1958), pp.126-149

DYNKIN, E.B. - Infinitesimal operators of Markov processes, Theory of Probability and its applications, 1 ; 1 (1956) (in Russian), pp.38-60

KAC, M. - On some connections between Probability Theory and differential and integral equations. Proc. of the Second Berkeley Symposium on Mathematical Statistics and Probability, 1951, pp.199-215

RAY, D.B. - Resolvents, Translation functions and strongly Markovian Processes, Ann. Math. 70, 1(1959), pp.43-72

RAY, D.B. - Stationary Markov processes with continuous paths, Trans. Amer. Math. Soc., 49(1951), pp.137-172

YOSIDA, K. - Functional Analysis, Springer-Verlag, 1978

Section 3

DOOB, J.L. - A probability approach to the heat equation, Trans. Amer. Math. Soc., 80, 1(1955), pp.216-280

DOOB, J.L. - Conditional Brownian motion and the boundary limits of harmonic functions, Bull. Soc. Math. France, 85(1957), pp.431-458

DOOB, J.L. - Probability methods applied to the first boundary value problem, Proc. of Third Berkeley Symposium on Math. Stat. and Probability, 2, pp.49-80

DOOB, J.L. - Probability theory and the first boundary value problem, Ill. Jour. Math., 2 - 1(1958), pp.19-36

DOOB, J.L. - Semi-martingales and subharmonic functions, Trans. Amer. Math. Soc., 77-1(1954), pp.86-121

HUNT, G.A. - Markov processes and potentials, Ill. Jour. Math., 1(1957), pp.44-93; pp.316-369 and 2(1958), pp.151-213

ITO, K. & McKEAN, H.P. - Potentials and the random walk, Ill. Jour. Math. 4-1(1960), pp.119-132

LEVY, P. - Processes Stochastique et Mouvement Brownien, Paris, 1948

Section 4

DOOB, J.L. - loc. cit.

FROSTMAN, O. - Potentiel d'equilibre et capacite des ensembles avec quelques applications a la théorie des fonctions, Thèse pour le doctorat, Lund, 1935, pp.1-118

ITO, K. - Stochastic Processes, Jap. Jour. Math., 18(1942), pp.261-301

LEVY, P. - Theorie de l'addition des variables aleatoires, 2nd ed., Paris, (1954) [Chap. 7]

RIESZ, M. - Integrale de Riemann - Liouville et potentiels, Acta Sci. Szeged, 9(1938); pp.1-42

Section 5

DOOB, J.L. - loc. cit. [Chap. 10]

ITO, K. - Stochastic differential equations, Memoir. Amer. Math. Soc., 4, 1951

Section 6

DYNKIN, E.B. - loc. cit.

FELLER, W. - The parabolic differential equations and the associated semi groups of transformations, Ann. Math. 55(1952), pp.468-519

FELLER, W. - On second order differential operators, Ann. Math. 61-1(1955), pp.90-108

FELLER, W. - On the intrinsic form for second order differential operators, Ill. Jour. Math. 2-1(1958), pp.1-18

RAY, D.B. - loc. cit.

TATA INSTITUTE OF FUNDAMENTAL RESEARCH

LECTURES ON MATHEMATICS AND PHYSICS

General Editors: S. RAGHAVAN and B.V. SREEKANTAN

MATHEMATICS

1. On the Riemann Zeta-function. By K. Chandrasekharan*
2. On Analytic Number Theory. By H. Rademacher*
3. On Siegel's Modular Functions. By H. Maass*
4. On Complex Analytic Manifolds. By L. Schwartz*
5. On the Algebraic Theory of Fields. By K.G. Ramanathan*
6. On Sheaf Theory. By C.H. Dowker*
7. On Quadratic Forms. By C.L. Siegel*
8. On Semi-group Theory and its Application to Cauchy's Problem in Partial Differential Equations. By K. Yosida*
9. On Modular Correspondences. By. M. Eichler*
10. On Elliptic Partial Differential Equations. By J.L. Lions*
11. On Mixed Problems in Partial Differential Equations and Representations of Semi-groups. By L. Schwartz*
12. On Measure Theory and Probability. By H.R. Pitt*
13. On the Theory of Functions of Several Complex Variables. By B. Malgrange
14. On Lie Groups and Representations of Locally Compact Groups. By F. Bruhat*
15. On Mean Periodic Functions. By J.P. Kahane*
16. On Approximation by Polynomials. By J.C. Burkill*
17. On Meromorphic Functions. By W.K. Hayman*
18. On the Theory of Algebraic Functions of One Variable. By M. Deuring*

---

*Not available.

19. On Potential Theory. By M. Brelot*

20. On Fibre Bundles and Differential Geometry. By J.L. Koszul*

21. On Topics in the Theory of Infinite Groups. By B.H. Neumann*

22. On Topics in Mean Periodic Functions and the Two-Radius Theorem. By J. Delsarte*

23. On Advanced Analytic Number Theory. By C.L. Siegel+

24. On Stochastic Processes. By K. Ito

25. On Exterior Differential Systems. By M. Kuranishi*

26. On Some Fixed Point Theorems of Functional Analysis. By F.F. Bonsall*

27. On Some Aspects of p-adic Analysis. By F. Bruhat*

28. On Riemann Matrices. By C.L. Siegel*

29. On Modular Functions of One Complex Variable. By H. Maass

30. On Unique Factorization Domains. By P. Samuel*

31. On The Fourteenth Problem of Hilbert. By M. Nagata

32. On Groups of Transformations. By J.L. Koszul*

33. On Geodesics in Riemannian Geometry. By M. Berger*

34. On Topics in Analysis. By Raghavan Narasimhan*

35. On Cauchy Problem. By S. Mizohata

36. On Old and New Results on Algebraic Curves. By P. Samuel*

37. On Minimal Models and Birational Transformations of Two Dimensional Schemes. By I.R. Shafarevich*

38. On Stratification of Complex Analytic Sets. By M.-H. Schwartz

39. On Levi Convexity of Complex Manifolds and Cohomology Vanishing Theorems. By E. Vesentini

40. On an Introduction to Grothendieck's Theory of the Fundamental Group. By J.P. Murre*

41. On Topics in Algebraic K-Theory. By H. Bass*

42. On the Singularities of the Three-Body Problem. By C.L. Siegel

---

+ Available as volume 9, T.I.F.R. Studies in Mathematics

43. On Polyhedral Topology. By J.R. Stallings*

44. On Introduction to Algebraic Topology. By G. de Rham

45. On Quadratic Jordan Algebras. By N. Jacobson*

46. On the Theorem of Browder and Novikov and Siebenmann's Thesis. By M.A. Kervaire

47. On Galois Cohomology of Classical Groups. By M. Kneser*

48. On Discrete Subgroups of Lie Groups. By G.D. Mostow

49. On the Finite Element Method. By Ph. Ciarlet

50. On Disintegration of Measures. By L. Schwartz

51. On Introduction to Moduli Problems and Orbit Spaces. By P.E. Newstead

52. On Numerical Methods for Time Dependent Equations - Application to Fluid Flow Problems. By P. Lascaux

53. On Optimisation - Theory and Algorithms. By J. Cea

54. On Deformations of Singularities. By M. Artin*

55. On Sieve Methods. By H.-E. Richert

56. On Irregularities of Distribution. By W.M. Schmidt

57. On Expansion Techniques in Algebraic Geometry. By S.S. Abhyankar

58. On Torus Embeddings and Applications. By T. Oda

59. On Forms of Higher Degree. By J.-I. Igusa

60. On Curves on Rational and Unirational Surfaces. By M. Miyanishi

61. On Wave Propagation. By G.B. Whitham

62. On Equations Defining Space Curves. By L. Szpiro

63. On Topics in Finite Element Solution of Elliptic Problems. By B. Mercier

64. On Diffusion Problems and Partial Differential Equations. By S.R.S. Varadhan

65. On Numerical Methods for Non-linear Variational Problems. By R. Glowinski

66. On Representations of Complex Semi-simple Lie Groups.
By T.J. Enright

67. On Nonlinear Waves and Shocks. By C.S. Morawetz

68. On Topics in Stochastic Differential Equations.
By D.W. Stroock

69. On Moduli of Curves. By D. Gieseker

70. On Partial Differential Equations. By G.B. Folland

71. On Three-dimensional Elasticity. By P.G. Ciarlet

72. On Sieve Methods and Prime Number Theory. By Y. Motohashi.

73. On Stochastic Differential Equations and Malliavin Calculus.
By S. Watanabe

74. On Results on Bezout's Theorem. By W. Vogel

75. On Stochastic Control and Nonlinear Filtering. By M.H.A. Davis